D1478817

PRINCIPLES OF **Electric Utility Engineering**

PRINCIPLES OF
Electric Utility Engineering

CHARLES A. POWEL

THE M.I.T. PRESS
Massachusetts Institute of Technology
Cambridge, Massachusetts, and London, England

Copyright, 1955
by
The Massachusetts Institute of Technology

All Rights Reserved

This book or any part thereof must not be reproduced in any form without the written permission of the publisher.

Third printing, January, 1963
Fourth printing, September, 1967

Library of Congress Catalog Card Number: 55-8181
Printed in the United States of America

PREFACE

During the years immediately following the Second World War, a large percentage of college students were veterans. They had had the opportunity of seeing for themselves to a greater or lesser degree the remarkable control and communication possibilities of vacuum tubes, magnetrons, computers, and all the other gadgets that make up "electronics." To them it appeared that there was little future in electrical engineering outside this fascinating field, and consequently the lack of interest in the generation, transmission, and application of sixty-cycle power which had already appeared as a trend before the war was greatly emphasized.

It was at this time that I was retiring after 43 years of activity in various phases of power engineering and was invited by Professor Harold L. Hazen, then Head of the Electrical Engineering Department, now Dean of the Graduate School, to the Massachusetts Institute of Technology to help revive interest in this class of engineering. It appeared that a course of lectures dealing with the everyday problems of the utility industry would be the best way of achieving the desired results. The text of this book is essentially the "notes" that formed the basis of the course, which covered two terms.

It is obvious that any chapter of my book can be expanded into several volumes. The book makes no pretense to be an exhaustive study of the subject, but on the contrary is presented as a concise review of those problems which in my long experience appear to me to be basic to the design and operation of a utility. It is intended as a guide to instructors wishing to give a similar course, whether in colleges or in utility training courses, and it should also prove of interest to those who, being employed in one branch of the industry, would like to get some idea of the problems met with in some other branch.

Although the material presented is not in any sense original, no attempt has been made to give credit to sources. Most of the discussion deals with principles so long established that it would be difficult to do so. Nevertheless, free use has been made of material published in such magazines as *Electrical World, Electric Light and*

Power, the journals of the engineering societies, and the publications of the various manufacturers. I wish to express my thanks for their helpful cooperation. In particular, I wish to thank Dean Frederick G. Fassett, Jr., for his patience in reading the manuscript and for his valuable suggestions in planning the final draft.

CHARLES A. POWEL

February, 1955

CONTENTS

Chapter 1 Corporate Organization, Objectives, and Finance . . 1
Franchise and monopoly. Standard of service. Organization. Physical property. Federal and state control. Utility definitions. Use of definitions and curves. Capital investment. General overhead. Fixed charges. Cost of money. Depreciation and obsolescence. Taxes and insurance. Production costs. Labor. Profit.

Chapter 2 Sources of Energy 24
Water power. Fuel. Standard installations. Gas turbines. Nuclear energy. Solar heat. Tidal power. Wind power. Gasoline from vegetation. Shale oil.

Chapter 3 Steam Generating Stations 33
Boiler plant. Control. Boiler rating. Steam turbines. Turbine rating. Condensers. Generators. Hydrogen cooling. Generator reactance. Inherent vibration. Exciters. Parallel operation. Selection of turbine generators.

Chapter 4 Generating-Station Auxiliaries 64
Auxiliary power. Fuel-handling equipment. Stokers. Fans. Pumps.

Chapter 5 Hydroelectric Generating Stations 79
Pump storage plants. Automatic stations. Types of water turbines. Governors. Generators. Ventilation. Rotors. Damper windings. Brakes. Auxiliaries. Combination steam and water power.

Chapter 6 Transmission of Energy 96
Frequency. Direct-current transmission. Industrial frequencies. Choice of voltage. Kelvin's law. Conductors. Corona. Conductor spacing. Cables. High-voltage cables. Cable failure.

Chapter 7 Transmission Equipment 111
Transformers. Switchgear. Circuit breakers. Types of breakers. Switchboards. Relays. Instrument transformers. Overcurrent relays. Differential relays. Directional relays. Distance relays.

Contents

Chapter 8 Power-System Fault Control 134
Supply circuits. Application of reactors. Application of circuit breakers. Examples of breaker applications. Phase segregation. Double-winding generators. Unbalanced faults. Application of relays. Machine faults. Transformer faults. Bus faults. Transmission-line faults. Grounding. System grounding. Generator grounding.

Chapter 9 Lightning Phenomena and Insulation Coordination . 160
Surge-measuring instruments. Stroke currents. Lightning protection. Standard wave form. Insulation coordination. Machine protection.

Chapter 10 Transmission Systems 181
Span and sag. Power transmitted. Conductor configuration. Counterpoises. Cost of transmission lines. Automatic circuit reclosing. Higher voltages.

Chapter 11 Power-System Stability 197
Power-angle diagram. Stability calculations. Network analyzer. Power circle diagram. Series capacitors. Unity power-factor operation. System loading schedules. Frequency control.

Chapter 12 Power Distribution 212
Subtransmission system. Primary system. Protection. Secondary system. Network distribution. Voltage regulation. Diversity. Distribution in commercial buildings. Distribution in industry.

Index 247

CHAPTER 1

Corporate Organization, Objectives, and Finance

Franchise and Monopoly

The electric utility business is unique in that what it sells—electricity—must be generated at the instant it is used. There is no possibility of setting up a stockpile to draw from in an emergency, however short. This simple fact has a dominant influence on every phase of the business.

Since the electric utility must of necessity utilize public streets and public property to bring its product to the public, it must obtain grants of public privilege, known as "franchises" from government. This privilege is usually granted only to a single concern in any given area, and thus a monopoly is created. The serving of an area by more than one utility is objectionable because the streets are cluttered up with facilities and plant and equipment are duplicated unnecessarily.

There are also economic reasons why it is in the public interest to have but a single utility in a given area. One reason is the high ratio of investment to revenue, reflecting high cost and heavy capital requirements, with the result that a relatively large portion of utility revenue is required to cover the return on investment or for compensation for cost of plant. Another reason is that the presence of a single facility results in more efficient use of plant and lower cost of service. The relative demands of customers for utility service occur at different times of the day and at different seasons of the year, so that the greater the number and variety of customers served by one utility system, the more efficient will be the use of the supply system, the cost of which is a large factor in the cost of the service. In technical language, increasing the total volume of business increases the diversity of demands of the various customers, the effect of which is to reduce the unit cost of the system capacity required. Increased volume of service also reduces the costs by making feasible the use of larger units of machinery and equipment, which in general have lower unit cost and better efficiency.

The electric utility is a monopoly in the sense that only one utility serves a given community. However, it does have to meet competition in the sense that, for instance, cooking, water heating, and domestic refrigeration can in many areas be done more economically by gas. In the industrial field there is always the possibility of putting in an isolated plant.

The franchise imposes on the utility certain duties, but it also gives it certain rights. The principal duties are that the utility:

1. Must serve all who apply for service.
2. Must serve up to maximum capacity and furthermore must be prepared to serve increases in demand beyond existing maximum capacity to the extent that such increases can be reasonably anticipated.
3. Must furnish service that is adequate and safe; must provide safe equipment through which to render service.
4. Must not discriminate unjustly among customers, but may serve different classes of customers under different rates.
5. May not obtain more than a reasonable price for its service.

The principal legal rights of a utility are that it:

1. May charge and obtain a reasonable rate for service. A reasonable rate is sufficient to recover all operating expenses, including a charge for depreciation of plant and a fair return on the investment "used and useful" in the public service.
2. May establish and enforce reasonable rules and regulations under which they provide service. These include such items as office hours for public contacts, prepayment discount, meter locations and inspections, service extensions and discontinuance of service.
3. May under certain circumstances invoke the law of "eminent domain" to obtain property.

Standard of Service

The service has to be continuous, 24 hours a day, 365 days a year. The cost of service is dependent on the amount of system capacity required to serve the load (demand) as well as on the volume (kilowatt-hours) used. The capacity connected to the system must at all times be sufficient to meet the demand.

From the viewpoint of system operation, the standard of service is determined chiefly by the following:

1. Reliability of service.

Corporate Organization, Objectives, and Finance

2. Quality of service.
3. Economy of production.

Reliability of service implies ability to supply electric energy to all customers whenever they demand it and in such quantities as may be required. Such reliability is measured by the number of interruptions to service and the number of consumers affected. It is dependent on sound design and competent operation.

Quality of service is indicated by the degree of attainment of standard voltage and frequency in the supply of electric energy. The design of the transmission and distribution systems determines how close to the standard the personnel can maintain voltage and frequency. Even competent personnel cannot maintain the standard if the system is poorly laid out.

Economy of production is effected through proper load division among the various stations on the system and among the equipment in each station.

Organization

The business organization of an electric utility will, of course, vary widely with its size, but the following may be considered typical of a medium-sized utility:

1. The board of directors.
2. Operating and staff organizations.
3. Engineering and construction departments.

The active management of the utility is in the hands of the president, as chief executive, and various vice presidents known as officers. The president is responsible for the general management of the company and for carrying out the broad policies laid down by the board of directors. He is always a member of the board, and in England is usually referred to as the managing director.

The relationship of the members of the board to the management of the company is purely advisory. However, they are not, as is widely thought, put there for decoration. The law imposes on the director the duty of giving to the affairs of the company the same degree of care that he would exercise in his own affairs. A director who habitually fails to attend the meetings of the board runs the risk of being sued for damages by the stockholders if the affairs of the company go wrong.

The officers of the company usually make up the top management. They are responsible to the president for the conduct of their respective operations. A typical list might include:

Principles of Electric Utility Engineering

1. Vice president in charge of operations.
2. Vice president in charge of sales.
3. Vice president and general attorney.
4. Treasurer and comptroller.
5. Secretary.
6. Director of personnel.
7. General purchasing agent.

These officers of course have a considerable number of persons under their supervision to keep the business running. Their function is fairly self-evident from their titles. For instance, the general purchasing agent is in charge of the procurement of equipment and supplies. He usually has on his staff one or more technically trained engineers with experience in operations to judge the suitability of the material being purchased. He is usually in charge of the store rooms and inventory.

The director of personnel is responsible for employee relations, employment, training, medical care, safety, pension administration, and so on.

The technical functions of engineering and operation may be under two officers or concentrated under one "vice president and chief engineer." The three principal subdivisions of this department are:

1. An operating manager responsible for:
    Generation.
    Transmission and distribution.
    System operation (chief load dispatcher).
2. A system planning engineer responsible for:
    System planning and load forecasting.
    Construction budgets.
    Technical research in system planning.
3. A manager in charge of engineering and construction responsible for:
    Station and line design.
    Construction.
    Engineering standards and specifications.

The operating manager is responsible for day-to-day operation of the system. His load dispatcher knows when and where the loads normally come on and go off and he must be prepared with the requisite generating capacity. This is not simply a matter of having a large amount of excess generating capacity running idle, because, as we shall see later, economy of operation is just as important in

Corporate Organization, Objectives, and Finance 5

this business as in any other. He must watch the weather. Sudden darkness causes the lighting load to increase very quickly, and thunderstorms may cause the loss of one or more lines.

The system planning engineer in contrast is responsible for the long-term plans. The delivery time of generating and transmission equipment will be anywhere from 6 months to 3 years. He must, therefore, anticipate the growth of the system in both time and location. This requires a careful study of civic, residential, and commercial developments, plans for roads and bridges, and so on. He uses as his tools population trend curves for the district, reports on business cycles, suggestions for development of natural resources, such as coal and petroleum, and so on.

The manager in charge of engineering and construction, except in very large utilities, will probably work in conjunction with some outside engineering firm as far as major construction is concerned. It usually does not pay to maintain an organization capable of carrying out all the engineering and construction work required by the utility. However, his department will probably lay out the broad outlines of all construction work, and in addition will do all the work required in substation and transmission line construction.

Physical Property

Physically the typical utility comprises the following:

1. One or more powerhouses. These may be thermal stations using coal, oil, or gas, or a combination of them as fuel, or they may be one or more hydroelectric stations, or they may be a combination of both types.

2. A "switchyard" for each power station. Where the power station is near a load center, the yard will have relatively low voltage distribution or subtransmission equipment. If the load center is at some distance the switchyard will contain transformers to raise the voltage to some suitable transmission value with their attendant circuit breakers and control equipment.

3. The transmission line itself with its towers or poles, usually erected on a private right-of-way.

4. The transmission substation with transformers to reduce the voltage to several subtransmission circuits with circuit breakers and control equipment.

5. The distribution substations where the subtransmission voltage is again reduced to supply the transformers mounted on poles for

6   Principles of Electric Utility Engineering

general distribution in suburban areas, or transformers in vaults for city underground distribution.

6. Substations to give service to large industrial customers at transmission voltage, at subtransmission voltage, or at distribution voltage, depending on the size and location of the factory and the most economical supply in the circumstances, the rates taking into account the type of equipment provided to give the service required.

In most cases the utility also undertakes to supply power to, and maintain, the street lighting systems. They usually also supply power to the street railways, where these exist. The utility responsibility usually ends at the house meter, or for industrial customers at the point of energy measurement.

Federal and State Control

The Federal Power Act, setting up the Federal Power Commission (FPC), places utilities in interstate commerce under the control of the FPC, and, under the rulings handed down by the Supreme Court, that means practically every utility in the country. The act reads in part:

"It is hereby declared that the business of transmitting and selling electric energy for ultimate distribution to the public is affected with a public interest, and that Federal regulation of matters relating to generation . . . and transmission . . . is necessary in the public interest . . . ."

"For the purpose of assuring an abundant supply of electric energy . . . with the greatest possible economy . . . the Commission is directed to divide the country into regional districts for the voluntary interconnection and coordination of facilities . . . ."

"No public utility shall sell, lease or otherwise dispose . . . of facilities subject to the jurisdiction of the Commission . . . in excess of $50,000 . . . or merge or acquire facilities . . . without an order of the Commission authorizing it to do so."

"The Commission may investigate and ascertain the actual legitimate cost of the property . . . . Every Public Utility . . . shall file with the Commission an inventory of all or any part of its property . . . ."

"Every licensee and public utility shall make, keep and preserve . . . such accounts, records of cost-accounting procedures . . . as the Commission may prescribe . . . ."

This act is the authority for setting up the Federal "Uniform System of Accounts" used by 95% of the utility systems. The Uniform System

Corporate Organization, Objectives, and Finance 7

sets up definitions of the various portions of the property as well as of the various accounts and is generally useful in making possible statistical comparisons of the utilities in different parts of the country. The utilities are divided up into four classifications.

Class A. Utilities having either (1) annual electric operating revenues of $750,000 or more, or (2) the original cost of whose electric plant amounts to $4,000,000 or more.

Class B. Utilities having annual electric operating revenues of more than $250,000 but less than $750,000, and the original cost of whose electric plant amounts to less than $4,000,000.

Class C. Utilities having annual electric operating revenues of more than $100,000 but not more than $250,000.

Class D. Utilities having annual electric operating revenues of more than $25,000 but not more than $100,000.

Class C and Class D utilities have a simpler system of accounts than Class A and B utilities.

"Transmission system" is defined as: (1) all land, conversion structures, and equipment employed at a primary source of supply (generating station or point of receipt of purchased power) to change the voltage or frequency for the purpose of its more efficient or convenient transmissions; (2) all land, structures, lines, switching and conversion stations, high-tension apparatus and their control and protective equipment between a generating or receiving point and the entrance to a distribution center or wholesale point; (3) all lines and equipment whose primary purpose is to augment, integrate, or tie together the sources of power supply.

"Distribution system" is defined as: all land, structures, conversion equipment, lines, line transformers, and all other facilities employed between the primary source of supply and of delivery to customers, which are not includable in transmission as defined above, whether such facilities are operated as part of a transmission system or as part of a distribution system.

All but two or three states also have public utility commissions that exercise control over the utilities. They have the final word on the rates to be charged and the return the utility is entitled to on its capital investment.

The purpose of all this control is to safeguard the public in the matter of rates. The rates which the utilities are permitted to charge are such that they will result in a "fair return on the fair value of the property." This was ruled by the Supreme Court many years ago. The "fair return" has been rather well agreed upon. The utility is entitled to charge rates which will enable it to:

1. Service its present capital obligations.
2. Provide an adequate return to the equity holders.
3. Attract new capital into the business to provide for expansion of facilities and replace obsolete equipment to adequately serve its customers.

The rates charged by the utilities vary widely in different parts of the country as many factors enter into the cost of generation and distribution, the principal ones being diversity factor, that is, the effective utilization of the equipment (capital investment) and the cost of fuel. The rates for the various classes of customers are published by the utility in a "rate schedule" which is available to the public. The customers are usually classified as residential, rural, commercial, industrial, public street and highway lighting, and other public authorities and utilities. The purpose of so classifying customers is to avoid discrimination in the application of the designated rates.

The residential and rural rates are simple energy charges, but the other rates are usually in two parts—a demand charge plus an energy charge. The demand charge is a fixed amount per kilowatt of maximum demand, usually ranging from $2.00 per kw per month for small consumers to $1.00 per kw per month for large consumers. The method of determining the maximum demand varies, but usually it is the greatest number of kilowatts averaged over any 15-minute period during any day of the calendar month. This may constitute the maximum demand on which the bill is based, or the highest maxima so determined may again be averaged over 2, 3, or 4 days. The demand charge is a return on the cost of the equipment installed by the utility to meet the demand, whereas the energy charge is intended to cover the operating and fuel costs. Since the utility is required by its franchise to stand ready at any time to provide the demand, the rate structure should rightly take this factor into account so that the price paid will reflect the true cost of the service.

Not many years ago this undoubtedly correct method of charging for electricity was also applied to residential customers in a half-hearted manner, rates being different depending on the number of rooms in the house, or the number of outlets, or the square feet of floor area. However, the general public could not be expected to understand these refinements, and general complaints led to their abolishment.

The money that these rates bring to the utility constitutes its gross income and pays for what it has to sell—electric energy. It is intended to cover all costs plus a fair return on the investment.

Corporate Organization, Objectives, and Finance 9

Utility Definitions

In order to provide uniformity in reports and accounts the terms used must be defined. Most of the definitions used by utilities were set up by the Industry and the American Standards Association prior to the passage of the Federal Power Act and were adopted by the Federal Power Commission. A few were added. The following are some of those most commonly used:

Capacity (FPC) is the load for which a machine, apparatus, station, or system is rated.

Capability (FPC) is the maximum load which a machine, apparatus, station, or system can carry under specified conditions for a given time interval.

Capacity factor (ASA) is the ratio of the average load on the machine or equipment for the period of time considered to the rating of the machine or equipment.

Plant factor (ASA) is the ratio of the average load on the plant for the period of time considered to the aggregate rating of all the generating equipment installed in the plant.

Demand (ASA) is the load at the terminals of an installation or system averaged over a specified interval of time. *Example:* Hourly kilowatt demand, seasonal kva demand.

Base load (ASA) is the minimum load over a given period of time.

Peak load (ASA) is the maximum load consumed or produced by a unit or group of units in a stated period of time. It may be the instantaneous load or the maximum average load over a designated interval of time.

Load curve (ASA) is the curve of power versus time showing a specific load for each unit of the period covered.

Load duration curve (ASA) is a curve showing the total time within a specified period during which the load equaled or exceeded the power value shown.

Load diversity (ASA) is the difference between the sum of the peaks of two or more individual loads and peak of the combined load.

Diversity factor (ASA) is the ratio of the sum of the individual maximum demands of the system to the maximum demand of the whole system.

Load factor (ASA) is the ratio of the average load over a designated period of time to the peak load occurring in that period.

Use of the Definitions and Curves

The best way of obtaining an understanding of the foregoing definitions is to assume a utility load curve and apply the definitions to it.

Principles of Electric Utility Engineering

The curve shown in Figure 1-1 represents the daily load curve of a system for 120 days. It would represent a system with a high ratio of household and commercial load to industrial load, resulting in a moderate morning peak and an evening peak of considerably higher value. More typical utility load curves are shown in Figures 1-2 and 1-3, but for purposes of illustration Figure 1-1 is more suitable.

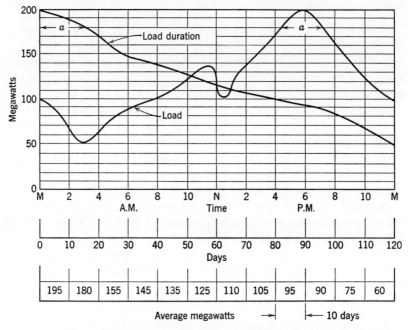

Figure 1-1. Daily load curve and load duration curve.

$$\text{Total megawatts} = \text{sum of average values}$$
$$= 1470$$
$$\text{Kilowatt-hours} = 1{,}470{,}000 \times 10 \text{ days} \times 24 \text{ hours}$$
$$= 353 \times 10^6$$

The load curve is necessary to the load dispatcher so that he can provide sufficient capacity on the bus to carry the load. It requires several hours to prepare a turbine generator to carry load. Therefore, the demand must be estimated in advance of its occurrence. The load curve gives a graphical representation of the demand and is usually drawn in the form of integrated hourly load for every hour of the year.

The variation of demand from hour to hour is not the same from day to day, but for any given day of the week (excepting holidays)

Corporate Organization, Objectives, and Finance 11

they usually show the same characteristic shape, with gradual seasonal changes. For instance, in summer the evening peak occurs later in the day than during the winter season. "Daylight-saving-time" usually creates a valley between 4:00 P.M. and 9:00 P.M. The duration of the evening peak is greater in winter than in summer. Public events of general interest broadcast over the radio are usually reflected in

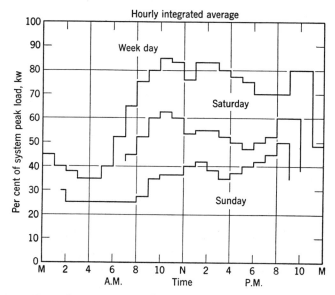

Figure 1-2. Typical summer week-day, Saturday, and Sunday loads of system in heavy industrial area.

the load curve, as are also general celebrations such as Christmas and New Year. Preparation must also be made for darkness brought on by storms during daylight hours.

Load duration curves are useful in studies involving:

1. The checking of computed station performance with actual performance.
2. The evaluation of difference of performance of equipment.
3. Transmission line economy studies.
4. Estimation of future production.
5. Estimation of water storage and run-off in hydroelectric developments.

The load duration curve is obtained from the load curve by using the same abscissa for the whole period under consideration as for the period represented by the load curve—or in the example 120 days and 1 day, respectively, on the assumption that the load curve of Figure 1-1 is typical of 120 days in the fall and wintertime. The annual load duration curve is obtained from the combination of typical curves for week days and week-ends at different seasons.

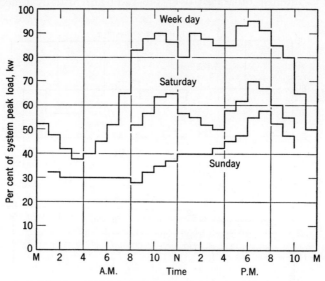

Figure 1-3. Typical winter week-day, Saturday, and Sunday load curves of system in heavy industrial area.

The total kilowatt-hours under the load duration curve must be equal to 120 times the kilowatt-hours under the load curve, but, since we have chosen a scale 120 times as great, the area under the two curves will actually be equal. This makes it easy to construct the load duration curve by merely plotting a curve which represents the length of time any given load between peak load and base load existed during the day. The average load during a small period of time (2 hours or 10 days in Figure 1-1) can be readily estimated, and the sum of these estimated values gives the total kilowatts.

In the example the load a (180,000 kw) lasts daily from 4:15 P.M. to 7:15 P.M. Transferred to the left-hand end of the seasonal scale this represents 15 days. The peak load, which is 200,000 kw, lasts about one hour a day. The base load is 50,000 kw. The average load for the 120 days under consideration is

$$\frac{353{,}000{,}000}{120 \times 24} = 123{,}000 \text{ kw}$$

Corporate Organization, Objectives, and Finance   13

If it is assumed that the installed capacity is 240,000 kw, the capacity factor for the period is

$$\frac{123,000}{240,000} = 51.3\%$$

The load factor is

$$\frac{123,000}{200,000} = 61.5\%$$

It is important to keep in mind the difference between "capacity factor" and "load factor."

The load factor is an indication of how good a job the sales department has done. If the customers were made up entirely of chemical

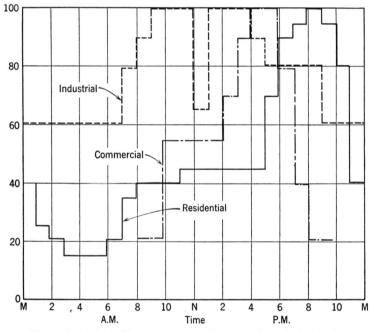

Figure 1-4. Probable composition of a typical week-day load.

and aluminum plants operating with a more or less constant load 24 hours a day, the load factor would be up in the nineties. If the load was entirely household load, we would get a sharp evening peak and a very poor load factor. The higher the load factor, the more economically the power stations can be operated. Practically all systems have one or two very efficient base-load stations that operate

14 Principles of Electric Utility Engineering

at constant load 24 hours a day. As we shall see, the steam economy of these stations is remarkably high.

The capacity factor is an indication of how good a job the operating and engineering departments have done. A high capacity factor means that the equipment is largely in continuous use and highly loaded.

At the end of 1952 there was installed in the United States a total generating capacity of 82 million kw, owned 80% by private investment and 20% by government. With this 82 million kw there was generated 400 billion kwhr. The average use of the installed capacity for the country as a whole was

$$\frac{400,000}{82} = 4880 \text{ hours}$$

The yearly capacity factor was therefore $4880/8760 = 55.7\%$. These figures are only roughly correct because some of the installed capacity was not in service for the whole year.

In 1938 the country generated 117 billion kwhr with 39 million kw installed, giving a capacity factor for the country as a whole of

$$\frac{117,000}{39 \times 8760} = 34.3\%$$

The large increase of 55.7% over 34.3% has not been altogether voluntary. A high capacity factor poses problems for the operators, who must have opportunities to clean boilers and make necessary repairs. What constitutes a desirable margin of capacity is a matter of opinion. It is sometimes expressed as 15% over peak load and sometimes by the rule that with the largest unit on the system out of service it should be possible to lose the largest unit in operation and still carry the peak. Such a rule, however, cannot be applied generally. Each system must make its own estimate of what reserve capacity is necessary to meet emergency conditions. It will depend on the size of the system, preponderance of industrial load over residential and commercial load, and to a large extent on what help the system can count on from its neighbors.

Diversity of demand and "diversity factor" stem from the fact that the coincident demand of a group of users is never as great as the sum of their individual demands. This is a most important economic factor in the utility industry, and diversity is a matter of continuous study. For instance, if four customers in different undertakings each have a peak load of 1000 kw, but it is known that the diversity factor in relation to those undertakings is 1.33, only 3000 kw of generating

Corporate Organization, Objectives, and Finance   15

capacity need be allotted to their load instead of 4000 kw. Diversity is discussed in more detail in Chapter 12.

Capital Investment

The investor-owned public utilities of the country as of 1952 had invested in their properties 25 billion dollars for a total of 65 million kw or $385 per kw. This amount represents the purchase of equipment over many years, much of it at a time when the dollar had greater value than today. However, the investment per kilowatt over the years has remained fairly constant because of the improvements in equipment, the larger size of the individual generators, transformers, etc., reflecting lower cost per kilowatt, and improved layout and utilization of power systems by the utilities.

In very general terms the capital of the typical utility can be considered as made up of 50% long-term debt, 15% preferred stock, and 35% common stock. On this investment a typical return will be 6%.

The items constituting the cost of running the utility business can be listed under the three conventional headings:

1. General Overhead:

General offices; executive and staff expenses; accounting; stores and warehouses; laboratories and repair shops; employee insurance, pensions, etc.; corporation taxes; legal and miscellaneous expenses.

2. Fixed Charges:

Cost of money; maintenance; depreciation and obsolescence; local taxes; insurance (fire, casualty, and liability).

3. Operating Costs:

Fuel; operating labor; water; miscellaneous supplies; superintendence clerical and testing.

General Overhead

As the name implies, the general overhead is that part of the fixed charges which is chargeable to the corporation as a whole and which cannot be applied to any particular part of it, such as power generation or distribution. The cost of the general offices, power sales, and legal and bookkeeping expenses are obviously such charges.

These general expenses tend to remain constant over fairly long periods of time even if radical changes take place in plant output, such as the addition of another power station or several substations.

There are also heavy corporation taxes apart from the local franchise and property taxes which must be spread over the whole operation. In some cases such taxes as "old age benefits," "sales taxes," and so

Principles of Electric Utility Engineering

forth are also imposed, and the total tax bill can easily reach as much as 20% of the operating revenue.

Fixed Charges

Overhead and fixed charges are essentially functions of the investment and are usually expressed as a percentage of the investment. It is obvious, therefore, that if the investment is known and the percentage to apply thereto, the evaluation of the items is simple.

The investment in power plants, which is usually expressed in dollars per name-plate kilowatt, varies widely, depending on location and the degree of development. There is a large item of expenditure in land, railroad and road connections, waterworks, offices, repair shops, cranes, etc., which must be made at the time the first unit is put in, but which changes very little, if at all, as further units are added. Therefore, the cost per kilowatt of a station in its initial stages is much higher than in its ultimate completion.

A survey of the literature indicates the cost of steam stations throughout the country built between 1927 and 1940 as ranging from $80 to $150, with a probable average of $110 per kilowatt based on the ultimate station capacity estimated at the time of building. Since 1940, equipment costs have gone up 50%, and construction costs have doubled.

The Mitchell Station of the West Penn Power Company containing two units of 80,000 kw cost $28,000,000 or $175 a kilowatt. The Public Service Electric Gas Company of New Jersey have stated that their new Sewaren Station with four units of 95,000 kw each will have cost $132 per kw when the fourth unit is completed. The Philips Station in Pittsburgh, with one unit of 60,000 kw, cost $184 per kw; the Mystic Station of the Boston Edison Company, with 150,000 kw installed, cost $155 per kw. It is seen from these figures that the utilities, construction engineering companies, and manufacturers have done a remarkable job of keeping down over-all costs of stations despite rising equipment and labor costs.

Cost of Money

To build the power plant, the utility will raise money by:

1. Selling stock in the corporation.
2. Selling bonds.
3. Borrowing from the bank.

A share of common stock, which may be of nominal par value ($10 or $100) or of no par value, is a statement of indebtedness with no

Corporate Organization, Objectives, and Finance 17

promise of a return or even of repayment. Dividends are paid on the stock from earnings of the company after all debts and taxes have been paid and interest has been paid on the bonds. The holder of corporation stock has no assurance that his original investment will remain intact. In practice it rarely remains constant; it either increases or decreases, depending on the fortunes of the corporation.

A bond on the other hand is a first lien on the property of the corporation. The corporation "promises to repay," sometimes on some definite date, and in the meantime undertakes to pay a definite annual return on the amount borrowed—usually $1000. Debentures are unsecured bonds.

The bonds are not usually sold direct to the public. They are purchased by a group of bankers known as "underwriters," who sell the bonds to the public at an advanced price, the difference being known as the "bond discount." If the public does not buy the bonds, the bankers have to hold them themselves. If an issue of 5% bonds running for 40 years is underwritten at 90 and sold by the underwriters to the public at 100, the company will have to pay every year $50 per $1000 bond. All it got, however, for its construction work was $900, so that the rate of interest is not 5% but $50 on $900 or 5.55%. Actually the money costs the company more than this because during the 40 years that it is paying interest the value of the bond is appreciating from $900 to $1000, or $2.50 a year.

At the end of the 40 years the principal of the bonds must be paid back. Money must therefore be set aside every year to build up a fund to do this. This is known as a "sinking fund," and it is built up from the depreciation accounts of the various pieces of property purchased with the proceeds of the loan. However, as we have seen, this is short of the required amount by the "bond discount." There must therefore be built up a supplementary account known as the "bond discount reserve." There will also be set aside from the income of the enterprise other reserve funds which legitimately come under "fixed charges." Over the years adjustments have to be made in the value of investments and inventories, and provision made for expansion. In other words, no enterprise can afford to consider all the money made over and above what they pay the bondholders as profit. A substantial part of the "profit" will be taken out and set aside as "reserves," which are usually plowed back into the business. What is left over is paid to the stockholders as "dividends." The "cost of money" expressed as a percentage will therefore be appreciably higher than the nominal rate of interest on the bonds, as will be shown in an example.

18 Principles of Electric Utility Engineering

The ratio of stock to bonds and to bank borrowing is of great concern to the financier. A corporation, if it is to prosper, must find investors willing to put up what is called "venture capital" since banks dealing with other people's money cannot lend all the money required. The amount they are permitted to advance on any kind of mortgage is prescribed by law. The investor will not put money into an enterprise which has a large prior debt, since his investment is thereby impaired. On the other hand, if the company is prosperous, the larger the bond issue the higher will be the return on the stock.

The cost of money varies over wide limits, depending upon financial conditions, the credit of the borrower, the popularity of the business, and the hazards involved. Municipalities can borrow money for power plants at interests as low as 3% or less. Private utilities borrow at interests of 3 to 6%, depending on their size and importance.

The low interest at which the government can borrow money is frequently given as the principal reason why utilities should be publicly owned. Actually the difference in this item as between a public body and an investor-owned utility is so small that, if all taxes and other costs are included by both classes of ownership, it has little influence on the rates.

Depreciation and Obsolescence

These items depend on the life of the plant which, in turn, depends on the design, materials, workmanship, and quality of the construction and equipment. Obsolescence depends, in addition, on changes in the state of the art. Old low-pressure steam turbines, which may have operated 10 or 15 years and which appeared to be near the end of their economic life, have been revitalized by the superposition of high-pressure units. A station containing two 30,000-kw, 400-psi, turbine generators may be equipped with new boilers and a 1250-psi turbine generator exhausting into the two old units, the three machines forming one cross-compound unit of, say, 80,000 kw. This procedure has been economically justified in a large number of cases, but obviously it complicates the depreciation problem.

Present experience indicates that the probable economic life of a steam plant is 25 to 35 years if considered as a whole. Good brick buildings have a probable life of 50 to 60 years; generating equipment boilers and auxiliaries 20 to 30 years; however, all these figures are conservative if considered in the light of actual practice today. A survey made in 1947 by the Federal Power Commission, covering two thirds of the generating capacity of the country, showed that a third of the generating capacity and a quarter of the boiler capacity in use then was over 20 years old. There were a few units still in operation over 35 years old, which indicates that old equipment remains operative even though its efficiency may be poor compared to similar modern equipment.

The method used in setting up a depreciation account is not universally the same among all utilities, but the trend is toward general adoption of the straight-line method. In 1948 already 73% of the utilities were using this method. The four methods used are:

1. Straight-line method.
2. Retirement reserve method.
3. Interest or sinking-fund method.
4. Revenue method.

The straight-line method provides for setting aside each year an equal proportional part of the cost of the property based on its estimated life. The "cost of the property" means the first cost of the equipment installed plus the cost of removal of the end of its life less the salvage value. For instance, a piece of machinery costing installed $100,000 (first cost less scrap value) and having an estimated life of 20 years would be depreciated at the rate of $5000 a year.

Retirement reserve is one that is built up by amounts credited to the account from time to time after inspection of the property, which in the judgment of the management will provide for current and future retirements.

The sinking-fund method is the same as the straight-line method except that the annual amount is calculated from compound interest formulas.

The revenue method builds up a reserve account by setting aside each year a percentage of the revenue. It has, therefore, no relation to the life expectancy or value of the equipment.

The purpose of the depreciation account is to set aside money to replace plant and equipment at the end of its useful life. For this purpose everything tangible owned by the utility is divided into "units of property" to be accounted for in accordance with the Fedeal Power Commission's "Uniform System of Accounts." It is defined therein as ". . . the loss in service value not restored by current maintenance, incurred in connection with the consumption or prospective retirement of electric plant in the course of service from causes which are known to be in current operation and against which the utility is not protected by insurance. Among the causes to be given consideration are wear and tear, decay, action of the elements, in-

20   Principles of Electric Utility Engineering

adequacy, obsolescence, changes in the art, changes in demand and requirements of public authorities."

The method of building up the depreciation fund is an accountant's problem and is of little interest to the engineer except insofar as he must use some depreciation percentage in evaluating alternative facilities. The amount set aside for depreciation annually is about 10% of the gross revenue, and the total amount in the depreciation reserve is some 20% of the capital investment.

Taxes and Insurance

The corporation taxes are included as part of the general overhead. In some cities the property used for public purposes is exempt from local taxes, but in most localities the utilities pay all the prevailing tax rates to the state, city, township, county or borough, in addition to Federal taxes. The aggregate tax bill is quite heavy and may amount to as much as 20% of the gross revenue. The local taxes are part of the fixed charges of generation in the given locality, as are also the fire and other insurance directly applicable to the local property.

Production Costs

The two large items in production costs are fuel and labor. Of the 400 billion kwhr generated in 1952, 286 billion came from private utility steam stations. To produce this energy the utilities consumed 106 million tons of coal, 63 million barrels of oil, and 887 billion cubic feet of gas.

The Btu content of a ton of coal will probably average 27 million; that of a barrel (42 gallons) of oil 7 million; that of a cubic foot of natural gas 1000. From this we deduce that in heat value 3.85 barrels of oil or 27,000 cubic feet of gas are equivalent to one ton of coal. Therefore, the equivalent tons of coal used by the power industry in 1952 was:

$$\begin{array}{r} 106 \times 10^6 \text{ tons of coal} \\ 16 \times 10^6 \text{ coal equivalent in oil} \\ \underline{33 \times 10^6 \text{ coal equivalent in gas}} \\ 155 \times 10^6 \text{ total equivalent coal} \end{array}$$

This is equivalent for the country as a whole to a coal consumption of 1.1 lb per kwhr, or a drop of 2 to 1 in 30 years.

Fuel consumption is given in pounds per kilowatt-hour at the station bus. With no losses, that is, 100% efficiency, the fuel consumption would be 3413 Btu per kwhr. Converting the above 1.1 lb of coal

Corporate Organization, Objectives, and Finance 21

to Btu at 13,500 Btu per lb (which is what we assumed), we obtain a consumption of 14,850, which corresponds to an efficiency of 3413/14,850 = 23%. This then was the over-all efficiency of all systems old and new in 1952 under the various load factors encountered in different parts of the country.

The efficiency of an individual station under test conditions will show a much higher percentage. Heat rates as low as 9286 Btu in modern high-pressure, high-temperature stations have been published during acceptance tests corresponding to 36.8% thermal efficiency. A base-load station will come close to attaining such an efficiency, but will not reach it because of inevitable falling off in capacity factor produced by heating up boilers, soot blowing, cleaning condensers, and necessary maintenance, all of which affect efficiency. The best systems must have stations which are not base-load stations. In one midwest system the four stations had load factors of 90, 70, 55, and 25%, the load factor for the system as a whole being 54%. It is obvious that the thermal efficiency of the four stations must have varied greatly.

Labor

The labor cost in the station is made up of direct labor and supervision. Direct operating labor is made up of two variables: (1) average wage paid and (2) output per man-hour, which depends on the efficiency of labor, type of fuel, type of plant, and to a large extent capacity factor.

In the last 10 years the payroll costs have doubled. As might be expected, the boiler room requires the most labor. As a rough estimate we can figure 50% of the total labor bill in the boiler room, 30% in the engine room and electrical bays, 15% engineering and supervision, and 5% miscellaneous.

The Federal Power Commission publishes labor and other production costs in mills per kilowatt-hour for all stations of Class A and Class B utilities. Table 1-1, taken from the publication, represents utilities which reported two or more stations with high and low plant factor. The results show the influence of plant factor on the cost of energy, but it should be remembered that the stations with low plant factor are probably the older and less efficient stations. The influence of plant factor comes from the fact that fuel and labor costs increase per kilowatt-hour with decrease in plant factor; the general efficiency is lower, and the auxiliary system demand does not drop much with load. Plant factor has a still more marked effect on the fixed charges, which go on whether the station produces or

Principles of Electric Utility Engineering

TABLE 1-1. ENERGY PRODUCTION COSTS

Mills per Kilowatt-hour

Area	Pacific Coast		Midwest		Midwest	
Plant Factor	69	44	89	51	60	47
Fuel	4.04	5.93	2.65	4.31	4.04	5.13
Labor	0.27	0.72	0.32	0.81	0.68	1.13
Water	0.01	0.06	0.04	0.11		0.02
Supplies	0.02	0.03		0.02	0.14	0.10
Maintenance	0.19	0.71	0.38	0.86	0.54	0.93
Total	4.53	7.45	3.39	6.11	5.40	7.31

Area	North Atlantic		South Atlantic		Northeast	
Plant Factor	100	67	75	53	82	44
Fuel	1.68	3.42	0.77	1.18	4.31	6.38
Labor	0.42	0.53	0.38	0.68	0.36	0.96
Water	0.02	0.02	0.01	0.02	0.01	0.03
Supplies	0.29	0.28	0.01	0.10	0.13	0.28
Maintenance	0.17	0.42	0.32	0.81	0.52	2.02
Total	2.58	4.67	1.49	2.79	5.33	9.56

not. The fixed charges are inversely proportional to the plant factor. A station operating at 100% plant factor one year and at 50% the next year will have fixed charges per kilowatt-hour the second year double those of the first year.

Profit

When all expenses have been covered and provided the business has been successfully operated, there remains a surplus which is in part put back into the enterprise for expansion and in part paid in the form of dividends to the owners of the property, that is, the shareholders. The last is what is usually referred to as profit. It is a reward for risks taken and for a management task well done. Without this profit there would be no incentive for people to invest their money in the enterprise, or for the management to feel personal responsibility for efficient operation.

As an example of what the financial picture of a utility looks like, the following example has been devised by averaging the 1952 financial reports of four representative utilities:

    I. Capital   493 million dollars
       Bonds   266 million dollars
       Preferred stock   67 million dollars
       Common stock   160 million dollars

Corporate Organization, Objectives, and Finance

II. Gross revenue   133 million dollars
    Operating expense   52 million dollars
    Taxes   26 million dollars
    Maintenance and depreciation   25 million dollars
    Interest on bonds   8 million dollars
    Dividends on preferred stock   3 million dollars
    Available for common stock   19 million dollars

III. Return on capital (millions omitted)
    $8 on $266 bonds   = 3%
      54% of capital in bonds   1.6%
    $3 on $67 preferred stock = 4.5%
      13.5% of capital in preferred stock   0.6%
    $19 on $160 common stock = 12%
      32.5% of capital in common stock   3.9%
                                                          6.1%

The return on the capital of this fictitious company is therefore 6.1%. A return of this order of magnitude is what is usually considered a "fair return" on the investment by most state commissions. It may go considerably higher for small utilities. The 12% return on the common stock does not necessarily mean that this amount was actually paid out in dividends. Some of it, as mentioned above, was retained in the business. The private utilities in recent years have been growing about 3 billion dollars annually. About two thirds of this amount comes from new capital, and about one third from reserves and surplus.

CHAPTER 2

Sources of Energy

All energy, even that which appears in some converted form, such as coal and oil, stems from the sun. Hence it may be said that all energy, except for that small amount which in the form of tides and wind may be due to the earth's rotation, is atomic at least in origin. The sun's energy is ascribed to the conversion of hydrogen into helium, two nuclei of hydrogen creating one nucleus of helium. The physicists have calculated that 564 million tons of hydrogen is converted every second into 560 million tons of helium, 4 million tons, or 0.7%, of mass being lost in the process, which according to Einstein's theory must appear in the form of energy. His formula is $E = Mc^2$, where E is the energy in ergs, M is the mass in grams, c is the velocity of light or 3×10^{10} cm per sec. The energy of conversion is, therefore, 79 million kwhr per lb, which gives us a truly fantastic figure as the total energy emanating from the sun.

The sources of energy for large-scale generation of electricity today are:

1. Steam from (a) coal, (b) oil, and (c) natural gas.
2. Water (hydroelectric stations).
3. Diesel power from oil.

Other possible sources of energy are direct solar heat, wind power, tidal power, shale oil, and atomic fission. None of these has as yet got beyond the pilot plant stage. However, as fossil fuels become scarcer and more costly, no doubt some of these, as well as petroleum manufactured from vegetable matter, may become useful and economical supplementary sources of power.

Water Power

Since the estimate of available water power depends on one's idea of how much of the theoretical flow can be harnessed, reckoning of the United States percentage of the world's water power resources

Sources of Energy

has ranged from 13 to 28%. One third of this water power is in the Pacific Coast States, one fourth in the Mountain States, and one fifth in the Southeastern States. It is anticipated that the Pacific Coast and the Mountain States will eventually have 30 million kw installed, but even so it is probable that the total water power capacity in the United States will continue to show a downward trend as a percentage of the country's total capacity. In the years 1947 to 1952 the installed water-power capacity decreased from 30 to 25% of the total.

Fuel

The supply of coal, oil, and gas in the United States is ample for many years to come. The estimated reserves of coal and lignite are about 3000 billion tons. This estimate is admittedly not much more than an educated guess, and is seriously questioned by many competent experts for the reason that many geologically inferred reserves may prove to be too costly to mine. Estimates of "recoverable coal" as low as one tenth this amount have been made.

The present consumption of coal in the United States is roughly 500 million tons a year, a figure which has been remarkably constant for some thirty years, except for the depression years. It is certain, therefore, that coal as a fuel will be available for many generations even on the basis of the lowest estimate of reserves.

Coal consumption by the utilities varies from year to year, depending on the relative costs of coal, oil, and gas. In recent years coal costs have risen sharply; consequently, while the use of coal in 10 years has risen 30%, the use of oil has gone up three times and that of gas 2.5 times. The marked increase in oil consumption has been in generating steam. The diesel engine plays a relatively small part in utility generation, the total installed capacity being under 2 million kw.

Standard Installations

The utilities in the United States, both privately and government financed, generate electricity from fuel and water power. Steam plants using coal, oil, and gas as fuel account for approximately 80% of the energy generated, and water power for 20%. A very small percentage of energy is produced from miscellaneous sources.

In all probability there will be no change in these more or less standardized types of steam and water-power plants in the foreseeable future except for inevitable improvements, but in their search for maximum economy the utilities are not overlooking other means of

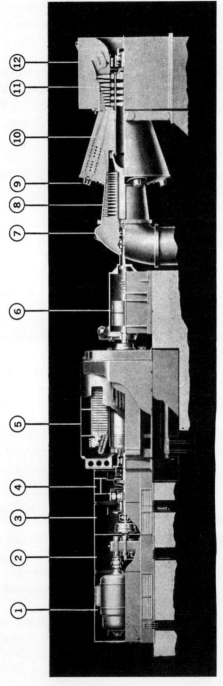

Figure 2-1. Principle of open-cycle gas turbine 5000-kw plant for power generation. (1) Starting motor. (2) Starting gear. (3) Disengaging coupling. (4) Exciter. (5) Generator. (6) Reduction gear. (7) Air inlet. (8) Air compressor. (9) Fuel nozzle. (10) Combustion chamber. (11) Gas turbine. (12) Exhaust. (Courtesy Westinghouse Electric Corporation.)

power generation. Of the many possibilities the following are at present receiving most attention.

Gas Turbines

Under favorable conditions the gas turbine can already compete economically with the diesel engine; but, whereas the diesel engine has already reached its maximum capabilities, the gas turbine shows possibilities of further development in both capacity and efficiency which may make it useful for peak load generation where cooling water is not readily available and fuel is expensive. Though manufacturers have been experimenting with gas turbines for 40 years, it is only in recent years that the machines have been available commercially. There are many installations of 3000 to 5000 kw, and several of 10,000 and 15,000 kw.

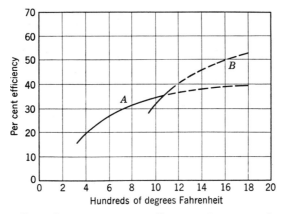

Figure 2-2. Effect of temperature on efficiency of steam and gas turbines. A. Modern large steam plant (projected). B. Expected best closed-cycle gas turbine plant. Efficiency of gas cycle is substantially unaffected by size of plant.

In principle the machine is simple. The plant (Figure 2-1) consists of a compressor, a combustor, a turbine generator, and a starting motor. The compressed air is mixed with fuel gas in the combustor and fed into the turbine. The excess power above what is required to drive the compressor drives the generator. Heat-recovery schemes such as the use of intercoolers in the air compression cycle, economizers in the turbine exhaust, and so on, are used to improve efficiency. The opencycle plant takes air from the atmosphere. If the turbine exhaust is connected to the compressor inlet a closed-cycle plant is obtained, which has a greater output-per-unit-weight than an open-cycle plant. For instance, if the absolute pressure of the working gas is multiplied

28 Principles of Electric Utility Engineering

by 10, the size of the plant theoretically is reduced by 10. Plants of this sort, which are not commercially available, would have the disadvantage of requiring large volumes of cooling water.

The appeal of the gas turbine lies in its promise rather than in its present performance. As we learn to make metals capable of standing higher temperatures, the gas turbine will take advantage of such developments better than steam turbines, as indicated in Figure 2-2, in which probable efficiency has been plotted against operating temperature. Nevertheless, even the present performance is such that the gas turbine plant is proving of great value in special cases. For instance, it has considerable appeal as a peak load plant for the reason that it can be started up from cold in a few minutes and can be located near the load center.

Nuclear Energy

A possible source of energy for power generation which has received much attention in recent years is derived from nuclear fission of uranium and thorium. These two elements lend themselves best to investigation in the present state of the art. The nucleus of uranium U^{235} is susceptible to slow neutron bombardment, whereby its mass is split into two smaller masses. In this process 1/1000 of the original mass is converted into energy, giving 11 million kwhr per lb of mass.[1]

Unfortunately most uranium is U^{238}, which is not fissionable. Natural uranium is made up of 140 parts U^{238} and one part U^{235}. There is a third isotope, U^{233}, which is negligible.

If we were dependent on U^{235} as the sole fuel for nuclear power plants it is questionable if they would ever be competitive with conventional steam plants. However, the physicists have developed a reactor, known as a "breeder reactor," which has economic promise. In this reactor, according to information made public by the Atomic Energy Commission, fissionable U^{235} is burnt in natural U^{238} to convert part of the latter to plutonium. Since plutonium is fissionable, the reactor generates more fissionable material than is consumed,

[1] Einstein formula: $E = Mc^2$ ergs

$$1 \text{ kwh} = 3.6 \times 10^{13} \text{ ergs}$$

$$1 \text{ lb} = 453.6 \text{ grams}$$

$$c = 3 \times 10^{10} \text{ cm per second} = \text{speed of light}$$

$$E = \frac{453.6(3 \times 10^{10})^2}{3.6 \times 10^{13}}$$

$$= 11.3 \times 10^9 \text{ kwhr per lb.}$$

Sources of Energy

and it is possible to convert all the uranium to fission products. In the process 140 times as much energy is released as when only U^{235} is consumed. This means that one has available as fuel relatively cheap and plentiful U^{238} in place of scarce and expensive U^{235}. At the same time a valuable by-product in the form of plutonium is obtained.

The reactor's power is generated as heat which is carried to a heat exchanger by a metal-alloy coolant to raise steam. The breeder reactor and heat exchanger thus replace the coal pile and boiler of a conventional steam plant.

Solar Heat

Since roughly one fifth of our fuel is used for heating buildings, a considerable load would be taken off our fuel demand if we could develop economically satisfactory means of utilizing solar heat for this purpose. Solar heating is well suited for this application because only low potential energy is required. The ordinary greenhouse illustrates the principle involved. The light and heat of the sun pass through the glass and are absorbed by the objects in the greenhouse. The objects re-radiate long infrared waves which cannot pass through the glass out of the greenhouse. The problem in solar heating of buildings is to store the heat for use at night and when the sun is not shining. It is said that with suitable material, such as Glauber salts, as a heat-storage medium, there would be no difficulty in storing enough heat to bridge over one or two weeks of sunless days. The architectural features and storage facilities required add so much to the building costs that solar heating is still uneconomical, but it holds considerable promise.

Tidal Power

At many places in the world high tides are encountered which represent a tremendous source of energy and, therefore, for a long time have interested engineers. Projects which have received serious consideration are the Mont St. Michel project in Brittany, the Severn Barrage in England, and particularly the Passamaquoddy project in the Bay of Fundy, which was thoroughly investigated by Dexter Cooper 25 to 30 years ago.

The high tides are due to resonant effects. If the length and depth of a bay are just right, the water in it will oscillate at its natural period, a balance being reached when the resistance forces and the disturbing force are equal. The disturbing force is, of course, the moon, which is responsible for all tides. Since these two forces are constant year in and year out, the tides and the power available can be predicted for any given time. This fact is a great advantage over river hydro developments, which vary from year to year over a wide range.

The original scheme proposed by Dexter Cooper must not be confused with the one on which later the Government spent much money. Cooper's scheme involved Canadian as well as United States waters.

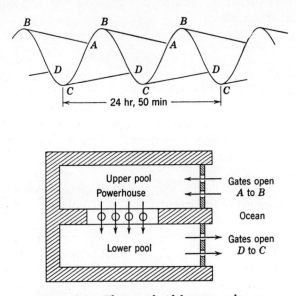

Figure 2-3. Three-pool tidal power scheme.

Fundamentally, what he proposed was a three-pool scheme, as shown in Figure 2-3. The tides vary in a sine wave twice every 24 hours and 50 minutes. Twice a day, therefore, when the rising tide reached A, the gates to the upper pool would be opened and the pool filled to B, when the gates would be closed. Twice a day when the tide fell to D the gates from the lower pool would be opened and the pool emptied to the point C, when they would be closed. The net head on the turbines would then be the difference between the upper line $ABAB$ and the lower line $CDCD$. The head varied from 16 to 25 ft. The theoretical available energy for the whole Bay of Fundy is 130 million kw, and at Quoddy Bay 2 million kw. Cooper expected to generate 500,000 hp. Using waterwheels of the variable-pitch propeller type, the station design presented no particular difficulties, but

Sources of Energy

there was considerable controversy regarding the feasibility of building some of the dams.

Under favorable circumstances (where a very large lower pool is available) use of a fourth pool makes a gain in energy possible. The fourth is an auxiliary pool at the station outlet within the lower pool. When the gates of the lower pool are closed at C, the water is permitted to collect in the auxiliary pool for 20 or 30 minutes and then is dumped into the lower pool, which in the meantime has been kept at the lowest level.

In the two-pool scheme proposed for the Severn Barrage energy is generated only during half the tidal cycle. Consequently a source of energy must be available during the other half cycle to make up the deficiency. For the Severn scheme, the British grid would have to make up this deficiency.

Wind Power

There are thousands of windmills the world over doing useful work, but they are all of small capacity. One exception was the 1000-kw experimental unit with a two-blade 175-ft wheel on a 125-ft tower installed at Grandpa's Knob in Vermont in 1941. It operated commercially for a short time, but was dismantled after a structural failure, as under war conditions repairs were not possible.

A new approach is being tried out in England, where a two-blade wheel 80 ft in diameter is expected to produce 100 kw. Instead of the wheel's driving the generator through gearing or by belt in the conventional manner, the blades are hollow with outlet openings at their tip. Centrifugal action forces the air inside the blades outward from a central hollow supporting tower. The air passing up the tower into the blades goes through a wind turbine, which drives a generator. The idea is illustrated in Figure 2-4.

There are many fundamental difficulties in building large-capacity windmills. Windmills of sufficiently rugged construction to withstand high winds are not capable of yielding net power at moderate winds. The power derivable from wind is proportional to the cube of its velocity. Since low velocities are useless and high velocities are dangerous, the limits of practicability are fairly narrow, of the order of 25 to 50 miles per hour.

Gasoline from Vegetation

Processes are already known for making alcohol and gasoline from vegetation, potatoes, sugar cane, or timber. Timber yields the maximum of carbohydrates per acre. It has been estimated that the 200 million tons a year of farm waste in the United States could be made to yield 90 gallons of alcohol per ton. There is a vast area of waste land in the tropics of South America, Africa, and Asia that could perhaps be made useful in producing alcohol from vegetation.

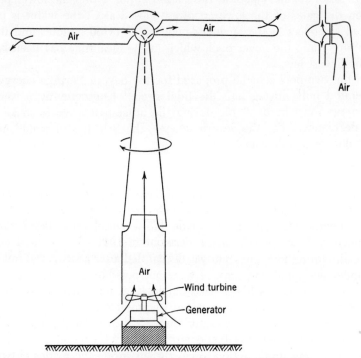

Figure 2-4. Principle of depression-type wind turbine.

Shale Oil

A great deal of interest is being shown in the production of oil from shale. It is estimated that there are shale oil reserves in Colorado and Utah to the extent of 340 billion barrels. The Bureau of Mines has already instituted experiments in the mining of the shale in a seam 70 ft thick which is said to produce 28 to 30 gallons of oil per ton of shale. The mining operation is reported as not difficult or expensive. The principal cost is in the extraction of the oil from the shale, for which no figure has as yet been published.

CHAPTER 3

Steam Generating Stations

The modern steam electric station dates from the installation of a 2000-kw steam turbine generator by the Hartford Electric Company in 1900. Up to that time power generation had been by reciprocating engines. Since the 5000-kw, 75-rpm engine generators at the Seventy-fourth Street powerhouse of the Manhattan Elevated Railway Company in New York stood 40 ft high and weighed 1 million lb, it is obvious that without the introduction of the steam turbine 5000 to 10,000 kw would be about the limit of generating units today.

With the introduction of the steam turbine, progress became rapid and continuous, and by 1925 single-shaft units had reached 50,000 kw in capacity at 1800 rpm. Today single-shaft units of 250,000 kw at 3600 rpm are being built. Since 1925 the Btu consumption of turbine generators per kilowatt-hour has been halved. Maximum throttle pressure has been raised from 1000 psi to 2400 psi, and the average for all stations from 350 psi to probably 1250 psi. In the same period the maximum throttle temperature was raised from 725 to 1100°F and the average from 675 to 975°.

A modern steam station consists of the following major equipment and auxiliaries:

1. Boiler plant
    Coal bunker
    Stokers or coal pulverizers
    Air preheaters
    Economizer
    Feedwater heaters
    Deaerator
    Boiler feed pump
    Forced draft fan
    Induced draft fan
2. Turbine generators
    Governor
    Generator cooling system
     (a) Air
     (b) Hydrogen
3. Condenser
    Condensate pump
    Circulating pump
4. Switchgear for generator control
5. Switchgear for auxiliary control

Figure 3-1 shows a cross-section through a modern steam station

34   Principles of Electric Utility Engineering

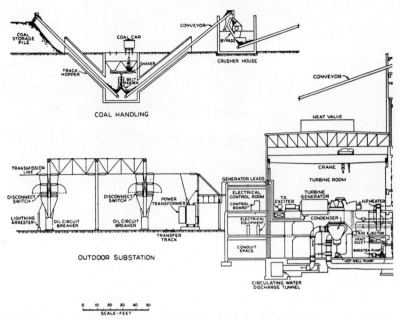

Figure 3-1. Cross-section through typical utility power station. (Courtesy of Stone and Webster Engineering Corporation.)

using powdered fuel. Where climatic conditions are favorable, outdoor and semi-outdoor stations are growing in popularity. Such a station is shown in Figure 3-2.

Boiler Plant

At least a superficial knowledge of the boiler plant is necessary for an understanding of the auxiliaries, which are for the most part electrically operated.

The first boiler of Watt's time was a water drum with a firebox immediately under it and a steam dome on top. Water was pumped into the bottom of the drum, and saturated steam at 30 or 40 psi was taken out of the dome. Later came Fairbairn's Lancashire boiler in which one or more fire tubes were inserted lengthwise through the water drum to increase the heating surface. Because of its simplicity and reliability, this boiler, with improvements and the addition of a superheater, mechanical stoker, and economizer, was used for many years, even after the introduction of water-tube boilers. The water-tube boiler as such is old, but general use of it developed rather slowly because of the difficulties with tubes. As materials improved, it gradually displaced all the other types in central-station service.

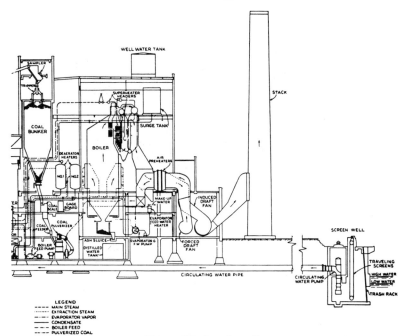

Figure 3-1 (continued).

All modern central-station boilers are of the high-pressure water-tube type, fired by pulverized coal, oil, or natural gas. There are three fundamental designs of water-tube boilers: the natural-circulation boiler in which the movement of the water in the tubes results from heat energy, the forced-circulation boiler in which the water is forced through the tubes by pumps, and the forced-flow or Benson boiler. The first two require a steam drum in which the separation of steam and water takes place. The advantage of forced circulation over natural circulation is that the boiler plant becomes considerably smaller, perhaps 30%. The forced-flow boiler is one in which the pressure is greater than the critical pressure of steam (3200 psi), at which the water is converted directly into steam without the intermediate latent heat stage. Thus there is no "separation" of steam and water, and there is no need for a steam drum. The principle of the three boilers is shown in Figure 3-3. A modern boiler is shown in Figure 3-4.

The boiler plant consists of the furnace, water walls, steam drum, superheater, economizer, air preheater, and a number of auxiliaries to be discussed in a later chapter.

The furnace, as the name implies, is the space in which the fuel is

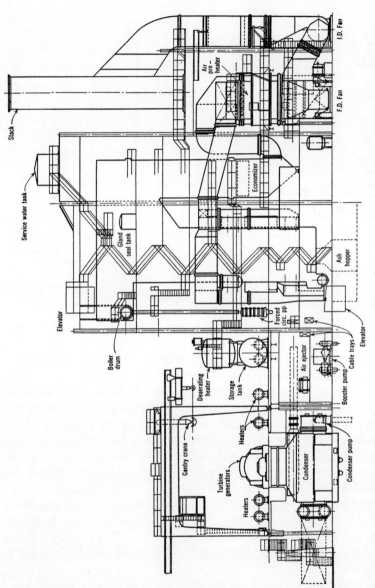

Figure 3-2. Cross-section through typical outdoor utility power station. (Courtesy of Stone and Webster Engineering Corporation.)

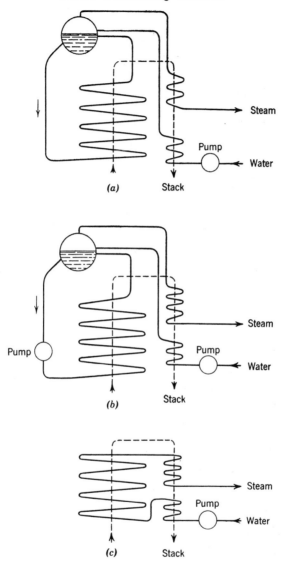

Figure 3-3. The three basic types of water-tube boilers.

burnt. In large boilers the fuel, powdered coal, oil, or gas is blown into the furnace through nozzles or special combustors. In small boilers (up to 350,000 lb of steam per hour) burning coal stokers are used.

The water walls are tubes in the boiler walls which serve the double purpose of cooling the walls and preheating the boiler feed-

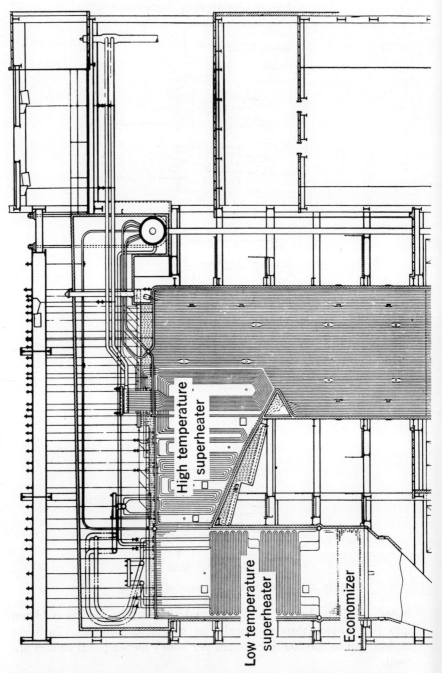

Figure 3-4. Modern utility steam-generating unit. One million pounds steam ment with secondary super-heater at outlet of one furnace and reheater

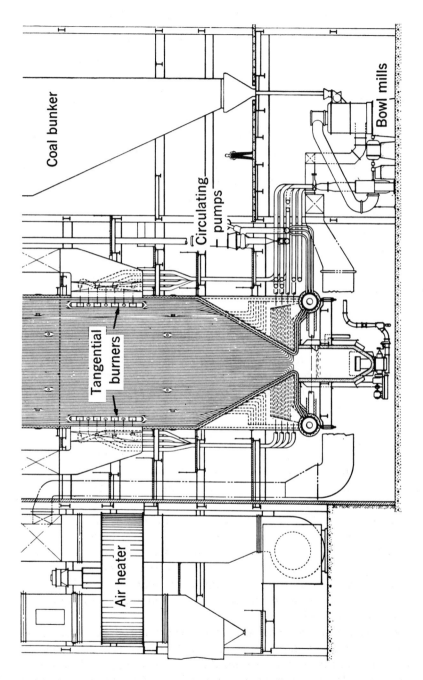

per hour, 1875 psi, 1000 degrees. The unit has separated furnace arrangeat outlet of the other. (Courtesy Combustion Engineering, Inc.)

40 Principles of Electric Utility Engineering

water. These water walls absorb by radiation an important part of the total heat released. Slag screens are water tubes above the furnace floor which serve to cool the molten fly ash or slag as it falls into the ash pit.

A superheater is a nest of tubes located in a hot part of the boiler not far from the furnace, through which the steam is passed on its way from the top of the drum to the steam turbine. It improves the economy of the heat cycle not only by increasing the Btu content of the steam, but also by reducing the moisture content of the steam. The control of the heat of the steam leaving the superheater is quite critical. It is accomplished by dampers or by "attemperators." The dampers merely by-pass more or less of the hot gases around the superheater. Attemperators reduce the temperature of the steam in the superheater by the injection of a small amount of feedwater.

The economizer is a heat exchanger through which the boiler feedwater passes. It is located in the coolest part of the furnace, or at the bottom of the stack to recover heat that otherwise would go up the stack. The economizer is not always used in modern steam stations.

The air preheater is located at the foot of the stack to extract heat from the flue gases and transfer it to the air used for the combustion of fuel. The greatest source of loss in a boiler installation is the quantity of heat carried away in the flue. Here is the reason for the economizer and air preheater to recover as much as possible of this heat.

The simplest heat cycle would be to pipe the steam from the superheater to the steam turbine, let it expand and do useful work in the turbine, condense the exhaust steam into water in the steam condenser, and feed the condensate back into the boiler. In actual practice the heat cycle is more complicated.

The over-all station economy can be improved several per cent by heating feedwater with partially expanded steam extracted from the steam turbine. Extraction of steam may take place in several locations along the turbine, providing steam at various pressures and temperatures for the progressive heating of the water and relieving the low-pressure stages of the turbine. A typical regenerative cycle is shown in Figure 3-5.

The steam thus extracted from the turbine is fed into "feedwater heaters." These are heat exchangers of the closed type in which the water to be heated and the extracted steam do not mix. They usually consist of a drum containing a nest of tubes. The steam is admitted to the drum surrounding the tubes through which the water flows.

Steam Generating Stations

One of the heaters, usually the last in the cycle, is a "deaerator." This, a heater in which the steam and water mix, is a device for removing air, oxygen, and other occluded gases from the water, since they tend to produce corrosion within the boilers, piping, heaters, etc.

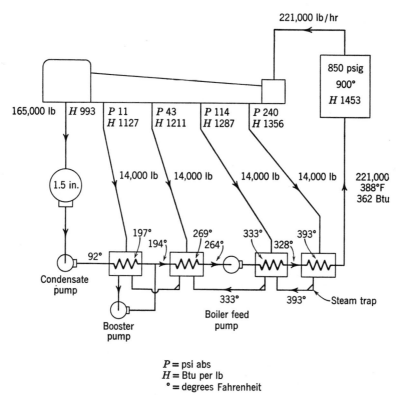

Figure 3-5. Example of regenerative feedwater heating for 25,000-kw turbine generator to obtain high over-all efficiency.

The number of heaters in the cycle varies with the initial steam pressure and the size of the turbine. The cost of the fuel also enters largely into the problem. Usually not more than four heaters are used, but under certain circumstances five, six, or even seven extraction points can be justified. The temperature drop and the volume of steam per heater should be approximately the same in all heaters for maximum efficiency.

Where fuel is expensive "reheat" can be justified. Reheat is a cycle in which the steam, after it has partially expanded in the turbine, is returned to the boiler and brought back to, or near, its original temperature. It is then piped back to the turbine to continue expansion through the lower pressure stages. By reheating, the energy available for work is increased, and the moisture in the lower stages of the turbine is reduced. A gain in efficiency up to 5% can be obtained through reheating, but the added complication in the way of piping and valves is considerable. Nevertheless, with the increase in the cost of fuel in recent years, the tendency has been to resort to reheat in the larger stations.

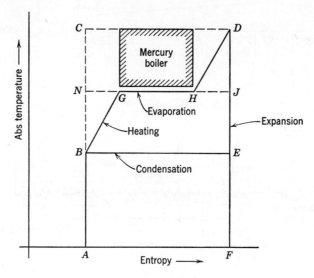

Figure 3-6. Binary cycle efficiency.

Another method of improving station efficiency is the use of the binary cycle, in which mercury is the high-temperature fluid. The General Electric Company has installed half a dozen such plants. In a binary cycle plant the mercury is vaporized at, say, 128 psi and 960°F in a suitable boiler. The mercury vapor passes through a turbine and leaves it at, say, 3 in. back pressure and 480°F temperature. The vapor is condensed in a "condenser-boiler" and in the process vaporizes water into steam at 450°F and 670 psi, which after passing through a superheater in the mercury boiler (825°F) drives a conventional steam turbine.

In the binary cycle the gain in efficiency is obtained from filling in the space above the steam evaporation line in the temperature-entropy diagram with useful work, as shown in Figure 3-6. In this figure the Carnot cycle efficiency is represented by the ratio of the area $BCDEB$

to the area $ACDFA$. It is evident that superheat represented by the line HD does not add much to the steam cycle (area $HDJH$), but it does add considerably to the Carnot cycle ($NCDJN$). The mercury turbine adds materially to the over-all mercury-steam cycle efficiency for the reason that, as evidenced by the diagram, it permits a much closer approach to the ideal Carnot cycle than is possible with superheated steam.

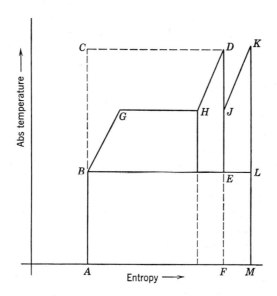

Figure 3-7. Reheat efficiency.

The reheat cycle does not permit the same gain. In this cycle (Figure 3-7) the steam first enters the turbine at D, expands to J in part of the turbine, is reheated to K, and then expands in the next stage of the turbine to L. The ratio of the area $EJKLE$ to the area $FJKMF$ represents the efficiency of the reheat cycle. If this efficiency is greater than that of the cycle without reheat (represented by the ratio of the area $BGHDEB$ to the area $ABGHDFA$), then there is a gain from reheat; otherwise not.

A large amount of space in the boiler house is devoted to treating the boiler feedwater. Even very small quantities of impurities in the boilers will quickly destroy the tubes. The water must be completely "soft," that is, free of calcium and magnesium. Passing the water through hydrated silicates of aluminum, so-called zeolites, which have the property of removing calcium and magnesium, softens it.

Control

Energy is fed into a boiler in the form of fuel and air. It is taken out in the form of steam to drive a turbine generator to supply an electrical load. If the speed and frequency of the generator are to be kept constant, the fuel, air, and water input into the boiler must follow closely the load variations on the generator. Hence automatic boiler control is necessary. There are many types of such control, and, though none can operate so quickly that load variations are followed exactly, for practical purposes the results obtained are satisfactory with some variation of steam pressure, up or down. The automatic control, which may be hydraulic, pneumatic, or electric, responds to variations in steam pressure and steam flow to regulate the rate of fuel and air supply. The proper proportioning of the two is very important and is achieved by measuring and comparing the flow of steam and the flow of combustion air. The volume of air required for complete combustion varies from 8 to 12 lb per lb of coal, depending on the heating value of the coal. It is closely proportional to the heating value and therefore is more or less constant per 1000 Btu. This fact permits steam flow to be used as a measure of fuel consumption, and the comparison of steam flow and air flow gives an indication of combustion efficiency.

The feedwater supply is controlled from a steam-flow meter, floats in the steam drum, or both.

Most new stations are equipped with devices for removing dust from the flue gases. These are available in two different types—mechanical and electrical. Mechanical collectors operate on the principle of forcing the flue gases into a whirling motion, whereby heavy particles are thrown out by centrifugal force and fall into dust hoppers while the gas itself passes up through a central core into the stack. Electrostatic precipitators utilize charged wires to attract and retain dust particles in the gases. The dust is periodically shaken into a dust hopper while the gas passes continuously into the stack. The use of these devices has been more or less required of utilities (and others using stacks) by public pressure. Without dust control a generating station of 250,000 kw might discharge several tons of dust per hour over the countryside, which of course would lead to complaints.

Boiler Rating

Boilers are rated according to their "normal steaming capacity for continuous operation and maximum steaming capacity" for a given period, usually one hour. The figures are given in pounds per hour,

Steam Generating Stations

and single boilers have been built with steaming rates of as much as 1,500,000 lb. The trend according to the boiler manufacturers is toward boiler capacities in excess of 1,000,000 lb, with pressures ranging from 1450 to 2400 psi and steam temperatures at the superheater outlet around 1050°F. However, higher pressures and temperatures are possible.

Steam Turbines

All steam turbines are based on two fundamental designs of blading—impulse and reaction. In impulse blading the expansion of steam takes place in the nozzle only, and the steam pressure is the same at the inlet and outlet edges of the rotor blades. In reaction blading expansion takes place in both the stationary and the moving blades, and the steam pressure is greater at the inlet than at the outlet edges of the blades. In the early days there used to be considerable argument as to their relative merits, but today all manufacturers use the type best suited to a given set of conditions, and combinations are found in a single turbine.

Central-station turbines on 60-cycle systems operate at 1800 rpm or at 3600 rpm. They fall into two classifications: single-cylinder and multi-cylinder.

In the single-cylinder turbine the steam expands from boiler pressure to condenser pressure in one cylinder. Such a turbine is shown in Figure 3-8a. It is always connected to a single generator.

In the multi-cylinder turbine the steam expands from boiler pressure to condenser pressure in two or more cylinders, which may be tandem-compound or cross-compound. Figures 3-8b and 3-9 show a two-cylinder tandem-compound turbine. Steam from the high-pressure cylinder exhausts into the low-pressure cylinder with double exhaust. In the final analysis, with a given maximum peripheral speed, it is the annulus formed by last row of blades which determines the output of the turbine. Theoretically each succeeding row of blades should be slightly longer than the preceding row, the increase in length following the adiabatic expansion of the steam. In practice the length of the last blade is limited by considerations of mechanical strength. At 3600 rpm the longest blades today are 26 in. long.

Figure 3-8c shows a tandem-compound three-cylinder double-flow turbine. The steam on leaving the first cylinder is returned to the boiler and reheated to its initial temperature (1000°F) before being fed into the second medium-pressure cylinder.

Cross-compound turbines also consist of a high-pressure cylinder and one or more low-pressure cylinders, but each is connected to its

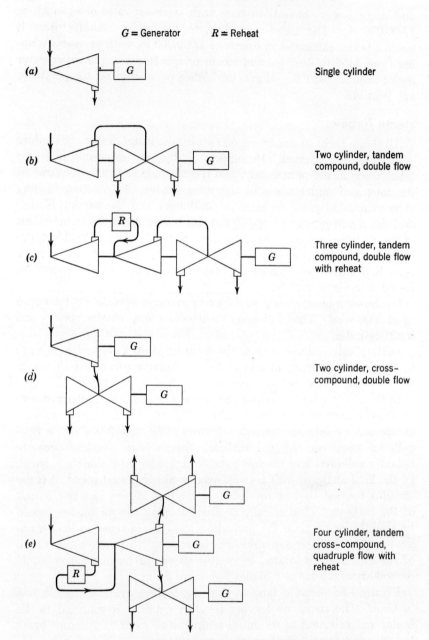

Figure 3-8. Typical arrangements of turbine cylinders.

Steam Generating Stations

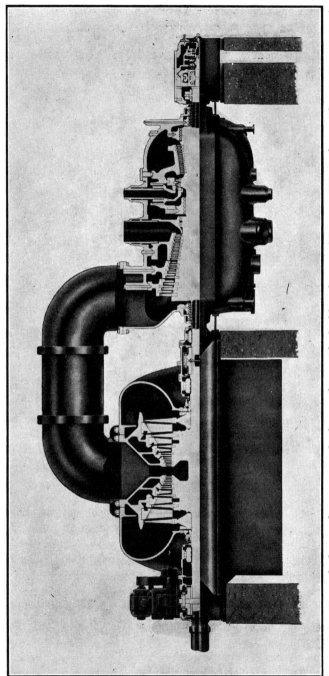

Figure 3-9. Two-cylinder, tandem-compound reheat turbine with double exhaust. (Courtesy Westinghouse Electric Corporation.)

48 Principles of Electric Utility Engineering

own generator, as shown in Figures 3-8d and 3-8e. Sometimes the high-pressure turbine generator runs at 3600 rpm and the low-pressure one at 1800 rpm.

The over-all economy, investment and operating costs, in any particular application dictates the choice of turbine-generator unit. The type of turbine depends on the size and steam conditions. In the

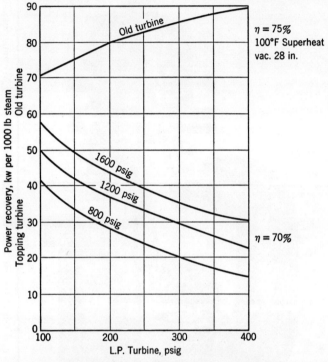

Figure 3-10. Power recovery from superposition.

matter of first-cost and space requirements a single-cylinder unit is given preference over a tandem unit, and a tandem unit over a cross-compound unit.

A method of cross-compounding is by "superposition," replacing old low-pressure boilers with a high-pressure boiler, the difference in the two pressures being utilized in a new turbine which exhausts into the existing turbines. Considerable capacity can be added to an old station in this manner at a reasonable cost and without requiring any additional condenser cooling water. This last factor alone may, in some cases, be the deciding factor in favor of superposition.

The curve in Figure 3-10 shows the gain that can be obtained from

Steam Generating Stations 49

superposition in kilowatt-hours per 1000 lb of steam. Assume that we desire to expand the output of an existing station containing two 20,000-kw units at 300 psi gauge, where no more condensing water is available. The output of the two units per 1000 lb of steam (according to the curve) is 86 kw, or a total steam consumption of 465,000 lb.

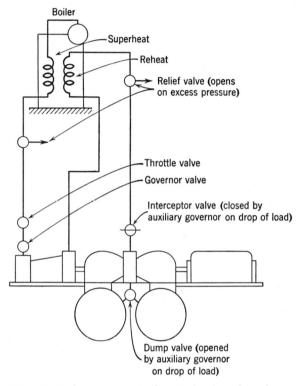

Figure 3-11. Typical arrangement of control valves for reheat turbines.

Replacing the boilers by a 1600-psi unit and adding a "topping" turbine generator allows an additional 36 kw per 1000 lb of steam to be obtained. The three machines must now be operated as a single cross-compound unit, with a 16,500-kw high-pressure element and two 20,000-kw low-pressure elements.

As stated before, reheat calls for a rather complex control system because the steam trapped in the reheat circuit between the governor valve, where the steam is shut off, and the re-entry point into the turbine may be sufficient to overspeed the turbine and trip the unit off the bus in case of a sudden drop in load. There must be provided special valves to by-pass the trapped steam into the condenser if there is loss of load. A typical arrangement of valves is shown in Figure 3-11.

The speed governors and lubricating systems are an inherent part of the turbine and are of no particular interest to electrical engineers, except insofar as the governor operating characteristic affects the stability of the system. This condition will be dealt with later.

Turbine Rating

The ASME and AIEE have jointly established a line of standard ratings for turbine generators up to 150,000 kw capacity. Concentrating on a few common ratings permitted the manufacturers to make savings in engineering and manufacturing costs, but chiefly it made possible much better delivery times through eliminating much drafting room work and making easier procurement of materials. Practically all units up to 150,000 kw that are sold today are built to these standard ratings.

The turbine rating is given in kilowatts at the generator terminals. In other words, a single rating is given for the over-all performance of the complete turbine generator. The performance of the unit is expressed as a "steam rate," that is, pounds of steam per kilowatt-hour, or as a "heat rate," Btu per kilowatt-hour. The steam rate is obtained from the formula $SR = \dfrac{3413}{(H_1 - H_2)NE}$, where $H_1 - H_2$ = available enthalpy in the steam Btu per pound, N = turbine efficiency, and E = generator efficiency.

For instance, the steam rate of a 30,000-kw turbine generator with 80% over-all efficiency operating at 850 psi, 900°F, exhausting into a condenser with 1.5 in. Hg back pressure, will be $\dfrac{3413}{(1458 - 900)0.8} = 7.7$ lb per kwhr.

Where steam extraction is used, the over-all station efficiency is improved, but for the same kilowatt-hour output more steam will have to be passed into the turbine. For example, the same 30,000-kw unit as a straight condensing machine will take in $7.7 \times 30,000 = 231,000$ lb of steam and exhaust the same amount. With three extraction points it will take in 261,000 lb and exhaust 198,000 lb. With four extraction points it will take in 263,000 lb and exhaust 196,000 lb.

Condensers

The condensers used in the larger central stations are invariably surface condensers in which the exhaust steam enters a cylindrical vessel from the turbine exhaust and passes over the surface of brass-

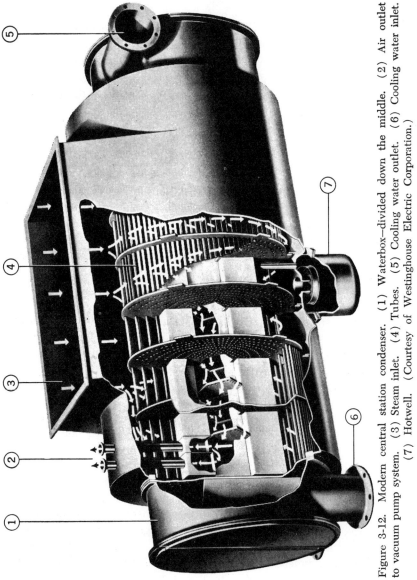

Figure 3-12. Modern central station condenser. (1) Waterbox–divided down the middle. (2) Air outlet to vacuum pump system. (3) Steam inlet. (4) Tubes. (5) Cooling water outlet. (6) Cooling water inlet. (7) Hotwell. (Courtesy of Westinghouse Electric Corporation.)

alloy tubes through which cold water is circulated. The back pressure in the condenser varies from 1 to 3 in. Hg, depending on its size and on the temperature of the cooling water. The vacuum is maintained by a "vacuum pump" which usually takes the form of an air ejector. In principle this is an expanding tube connected to the air take-off of the condenser. Steam is admitted to the ejector through a small nozzle in its neck, and, expanding through the ejector, entrains the air and gas with it into a small so-called after-condenser and out into a drain. In most cases two-stage ejectors are used, the first stage removing the air from the main condenser into an intermediate condenser, and the second stage taking it from the intermediate condenser into the after-condenser.

The condensed steam in the condenser shell falls into the "'hotwell," the sump at the bottom of the condenser shell in which the condensed steam ("condensate") collects, and from which it is pumped by the condensate pump into the first-stage heater. The condensate pump is usually a two-stage centrifugal pump.

At either end of the condenser shell are the water boxes into which the circulating water is pumped. In modern large-size condensers the water boxes are divided vertically in two parts, each half having its own circulating pump, so that one half of the tubes can be cleaned while the other half is still in use. The loss of vacuum when one half of the condenser is out of service is not enough to reduce the output of the turbine seriously. Figure 3-12 illustrates a modern condenser.

A modern high-pressure steam station requires about one gallon of cooling water per minute per kilowatt rating; thus for normal conditions a 60,000-kw unit requires about 60,000 gallons of condensing water per minute. In small units this figure may go up to 2 gallons per kilowatt per minute. A variation in vacuum from standard conditions of one inch up or down will change the steam consumption 5 or 6%.

Generators

All central-station generators are two-pole or four-pole machines (3600 or 1800 rpm at 60 cycles[1]), direct-connected to the steam turbines. In cross-compound units two or more generators constitute one "unit," and they are treated electrically as a single machine. The generators are connected to a stub bus usually through disconnects, the stub bus being connected to the station bus through a circuit breaker. When the unit is started up the two generators are excited with no-load normal-voltage field current from the spare exciter motor

[1] See Chapter 6 for a discussion of system frequencies.

Steam Generating Stations

generator set before steam is supplied to the turbine. Both machines are brought up to speed, and, since they have excitation, they remain locked in step. The two machines are then synchronized with the station bus around the circuit breaker.

Generators are rated in kilowatts and kva at a certain power factor and short-circuit ratio. Two-pole (3600-rpm) machines are being built up to 250,000 kva rating. Up to 150,000 kw the ratings have been standardized by a joint committee of ASME and AIEE. The smallest standard generator, 12,650-kw, is air-cooled, and all the larger ones are hydrogen-cooled.

The rating and dimensions of generators are affected by power factor, short-circuit ratio, and hydrogen pressure. The standard generators are rated at 0.85 power factor, 0.8 short-circuit ratio, and 0.5-psi hydrogen pressure.

The power factor is the ratio of the real power to the apparent power, kilowatts to total kva. The standard generators will not give full kva output at any power factor. As the power factor decreases, the total kva also decreases because of limitations in the field; for instance, at 75% power factor the 30,000-kw generator gives only 32,000 kva instead of its normal rating of 35,000 kva. At zero per cent power factor the machine gives 25,000 kva.

The armature flux can be considered as made up of two parts—the flux of armature reaction and the flux of armature reactance. The flux of armature reaction is the main part of the armature winding flux which combines with and modifies the main flux. It links with the field coils and rotates with them. The flux responsible for armature reactance is that minor part of the armature flux that does not penetrate the field coils, such as that surrounding the armature coil ends. To obtain a mental picture of this phenomenon imagine the armature coils as ordinary reactance coils in which a current is flowing. Actually if full load current were to circulate in the armature coils with the machine at standstill, reactance drop would be obtained. It would, however, be incorrect because of the presence of the field structure. There is no way of separating the two flux components by test, but it can be done graphically from the saturation curves. It is the armature reaction that causes the kva output of the generator to fall off with power factor. At zero power factor armature reaction subtracts directly from the main excitation; a much greater current is therefore required in the field coils to overcome this reaction and maintain voltage, and this current produces excessive heating in the field structure.

The short-circuit ratio is the ratio of the field current required to generate rated voltage at no-load to the field current required to

54 Principles of Electric Utility Engineering

circulate rated armature current through the short-circuit armature (both at rated frequency). The effect of short-circuit ratio on generator rating is not so obvious as that of power factor. In operation mechanical energy is transferred from the rotor to the stator by means

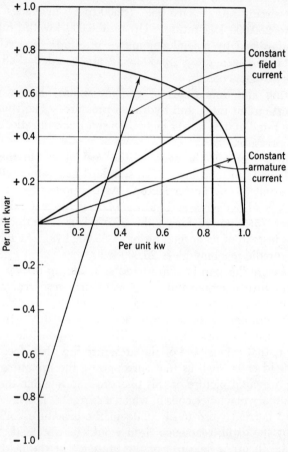

Figure 3-13. Output of standard turbine generators rated 0.85 power factor, 0.8 SCR.

of the magnetic flux in the air gap. This flux depends on the magnetizing effect of the exciter and the demagnetizing effect of the armature current. The greater the magnetizing effect as compared to the demagnetizing effect, the higher the short-circuit ratio, and also the greater the physical dimensions of the machine to provide the greater net flux. In other words, for a given physical size, the lower the short-circuit ratio the higher the generator can be rated.

Steam Generating Stations

There are thus two inherent limitations to the output of generators in operation; the full-load armature current for which the machine is designed, and the maximum field current the rotor will carry without overheating. The curves in Figure 3-13 show these limits for the standard ASME-AIEE generators. The arc centering about the origin represents the limit imposed by constant armature current. The other arc is the limit imposed by constant field. This arc is obtained through considering the generator as a simple transmission line with constant reactance equal to the synchronous reactance of the machine X_d, and constant receiving and sending-end voltages equal respectively to the terminal voltage e_t and the internal voltage e_d. The power circle of such a transmission line will have its center on the negative reactive axis at e_t^2/x_d. Since with no saturation $1/x_d = SCR$, the center is found through marking off the short-circuit ratio value on the negative reactive axis. Its radius must pass through the point of rated real power and rated reactive power P. With a value of short-circuit ratio of 0.8 a generator designed for 0.85 power factor will give at zero power factor about 75% of its rated kva.

The purpose of a high short-circuit ratio is to keep the machine in synchronism with the system during a disturbance. In American practice it is unusual to use a short-circuit ratio of less than 0.8. However, with the much improved voltage-regulating systems available today and the possibility of quickly building up the excitation following a disturbance, there is some opinion that a lower short-circuit ratio might prove entirely satisfactory. A lower short-circuit ratio permits a reduction in machine size for a given kilowatt output—a very important consideration where capacities of 250,000 and 300,000 kw are being considered.

Hydrogen Cooling

All steam-driven turbine generators above 15,000 kva capacity are hydrogen-cooled. This requires that they be built with gas-tight housings and certain control equipment to prevent explosion, but the advantages of hydrogen cooling in greater output per unit weight are such that the extra complication is fully justified. The effect of hydrogen cooling on the rating is to reduce the losses for a given temperature rise. Because of the lower density of hydrogen, windage losses in the generator are one tenth of their value in air. Hydrogen has a thermal conductivity seven times that of air, which reduces the thermal resistance in all heat-flow paths. Also the rate of surface-heat transfer in hydrogen is 35% greater than in air. Because of these advantages greater output can be obtained from a given amount of active material

Principles of Electric Utility Engineering

for the same temperature rise. The effect of corona is negligible in hydrogen, insulation retains its flexibility longer in the presence of hydrogen, and finally, because hydrogen will not support combustion, fire hazard is reduced.

In order to be explosive the percentage of hydrogen in the hydrogen-air mixture must lie between 5 and 75%, and the control equipment is designed to operate an alarm if the purity of the hydrogen in the generator housing drops below 95%. In the 20 years that these machines have been used there has been no explosion. To prevent danger while the machine is being charged, or on any occasion, such as inspection, when a change is being made from air to hydrogen or from hydrogen to air, we must first fill the machine with an inert gas (carbon dioxide) and then replace the inert gas with hydrogen. We must never permit air and hydrogen to come together in the machine.

The hydrogen-control equipment consists of a number of hydrogen bottles connected to a header, which in turn is connected to the generator through a pressure valve and pressure regulator. The density meter and alarm system is in principle a small constant speed blower circulating the mixture. If the density of the mixture varies because of the presence of air, the drop of pressure across the fan will also vary and show up on the meter.

In the past the machines have been designed for operation at gas pressures of 0.5 to 15 psi. Experience has shown that as the pressure is increased from 0 psi to 15 psi gauge, the output of generators of conventional design can be increased 1% for each pound increase in the gas pressure without exceeding the temperature rise guarantee at atmospheric pressure. Over a range of 15 to 30 psi the increase in output is about 0.6 of 1% increase in pressure. These percentage increases in output are based on 100% rating at 0.5 psi gas pressure. As we go higher with the pressure, the gains fall off, so that 30 to 50 psi would seem to be an optimum pressure. This is because the heat from the copper conductors must pass through the coil insulation before being carried away by the hydrogen gas. In recent designs the hydrogen is brought into more intimate contact with the copper conductors, where the heat is generated and carried away directly. Here advantage can be taken of the higher pressures—45 to 90 psi.

One such design is the so-called inner-cooled generator. The rotor conductors are hollow, and high-pressure hydrogen passes through them. In the stator the conventional Roebel transposed copper strands are used, but between them a thin-walled copper tube runs the whole length of the slot, and hydrogen is forced through it. Thus in both the rotor and the stator the hydrogen removes the heat directly from

Steam Generating Stations

the copper instead of through the insulation. The output per pound of active material is more or less doubled by this method of cooling, and two-pole generators up to 300,000 kva can be shipped assembled, as against 150,000 kva with conventional designs.

Before operation at a greater pressure than 15 psi, state codes applying to "unfired pressure vessels" should be studied. Some of these codes require annual inspection and hydrostatic tests which are quite stringent.

Generator Reactance

Depending on the type of problem being dealt with, three values of generator reactance are of interest in application-engineering work. When a generator is short-circuited the initial current is high and is determined by the "subtransient reactance" of the machine. In a very few cycles the current sinks to a lower value determined by the "transient reactance," and after 30 to 120 cycles it settles down to a final value determined by the "synchronous reactance." The synchronous reactance is therefore that value of reactance which the machine has after all transient effects due to change in armature current have died out. It can be obtained from the no-load saturation curve and the short-circuit saturation curve as shown in Figure 3-14. Since there is no load on the generator, the terminal voltage may be assumed equal to the induced voltage E. With the armature short-circuited, when the terminal voltage must of necessity be zero, the voltage E at any field value I_f must therefore be entirely consumed in forcing the current A at the same excitation value through the impedance of the armature, or $X_d = E/A$, neglecting resistance and saturation.

If normal voltage on the no-load, open-circuit, saturation curve is expressed as 100% E, with corresponding field amperes 100% I_f, and full-load current on the short-circuit saturation curve is expressed as 100% I, with corresponding field amperes I_{fs}, then the voltage corresponding to I_{fs} on the open-circuit saturation curve represents the voltage drop through the short-circuited armature at full-load current due to synchronous reactance. The curve shows that this voltage drop expressed as a percentage of normal voltage is equivalent to field current corresponding to rated current divided by field current corresponding to rated voltage. It will be seen from the definition previously given that this is the reciprocal of short-circuit ratio. In other words, as an approximation synchronous reactance $X_d = 1/SCR$.

The synchronous reactance is the value that would establish the regulation of a generator, or the voltage drop with added load, if it had no voltage regulator. Most generators are provided with an automatic voltage regulator which with variations in load restores the terminal voltage to normal within a more or less short period of time. Regulation is usually of importance only in those cases where the load change may be large in relation to generating capacity,

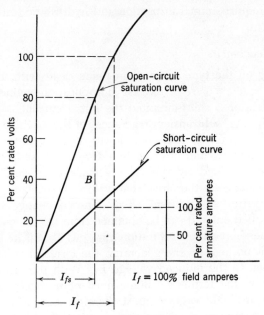

Figure 3-14. Relation between synchronous reactance and short-circuit ratio.

$$\frac{100}{B} = \frac{I_f}{I_{fs}} = X_d = \text{synchronous reactance}$$

By definition $\quad \dfrac{I_{fs}}{I_f} = SCR = \text{short-circuit ratio}$

as in a factory with its own generating plant supplying large motors. When such a large motor is started the voltage drop may be sufficient to cause other motors to stall. It is important in such cases that a quick-acting voltage regulator be used.

Inherent Vibration

Large turbo-generators inherently have a double-frequency stator vibration due to the magnetic pull of the rotor on the inside of the stator core. In such machines, therefore, the stator iron is not rigidly connected to the outer frame, but is attached thereto through

Steam Generating Stations

springs or links which permit the inner core to vibrate without communicating the vibrations to the foundations. There is also a double frequency vibration of the rotor because of the physical dissymmetry resulting from the two-pole winding slots. In large machines with long rotors where this vibration might become objectionable, narrow slots perpendicular to the slots are machined in the pole faces to equalize the moments of inertia in the two principal axes.

Exciters

Some 25 to 30 years ago when engineers became interested in system stability, it was recognized that the performance of the excitation system was a major factor in the problem. System stability is the ability of several power stations connected through transmission lines to remain in synchronism during system disturbances. It was realized that if the excitation of the generators could be increased quickly, there was a possibility of counteracting the effect of armature demagnetization. Thus came into being what is known as "quick response excitation."

When a fault occurs causing a heavy current suddenly to flow in the armature of a synchronous machine, there is an immediate drop in terminal voltage due to the leakage reactance drop. There is then a further drop in the terminal voltage due to the armature demagnetization, the effect of which is to reduce the air-gap flux producing the induced voltage. This second phenomenon is relatively slow and with quick response excitation it can be counterbalanced to a large extent.

The method used to speed up excitation was to:

1. Build the exciter with a low time constant by subdividing the field into several parallel paths instead of connecting all pole windings in series as had been common practice until then.

2. Keep a constant voltage on the exciter field by means of a "pilot exciter" instead of making the machine self-excited. The pilot exciter had the further advantage of making possible the elimination of the bulky and costly "main field rheostat" in the alternator field circuit.

3. Make the voltage regulator responsive to all three phases instead of one phase as heretofore. When a single-phase fault occurs on a generator, it can happen that the voltage of one phase will actually rise while the voltage of the other two phases drops. If the single-phase regulator happens to be connected to that phase, it will operate

60   Principles of Electric Utility Engineering

to decrease excitation instead of increasing it, thus from the stability point of view doing more harm than good.

4. Arrange the circuit so as to add or remove resistance in the exciter field circuit quickly in case of a major disturbance to change the voltage applied to the field windings. For this purpose the voltage regulator was equipped with contacts for inserting or short-circuiting a block of resistance in addition to the contacts that set in motion the motor-operated rheostat to take care of small changes in voltage. The principle is shown in Figure 3-15.

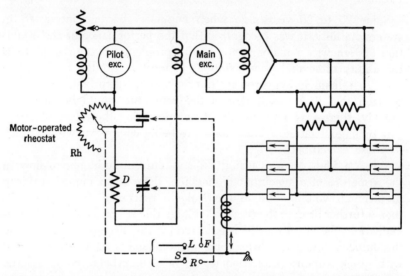

Figure 3-15. Quick-response voltage regulator. $L-R-S$ = lower and raise contacts for minor voltage changes. $L-R-F$ = field forcing contacts for large variations in voltage. LF opens up contacts around resistor D, and RF short-circuits the main field rheostat Rh.

It is, of course, possible to build exciters for various speeds of response. It became necessary, therefore, to set up a definition of what was meant by quick response, and the ASA has such a definition. Nominal exciter response is defined as the numerical value obtained when the nominal collector ring voltage is divided into the slope, in volts per second, of that straight-line voltage-time curve which begins at nominal collector ring voltage and continues for one half second, under which the area is the same as the area under the build-up curve of the exciter starting at the same initial voltage and continuing for the same length of time.

Figure 3-16 illustrates this definition. Most exciters are built with

Steam Generating Stations

a speed of response, as defined, between 0.5 and 1.0. Speed of response has lost much of its importance with the advent of high-speed breakers. The reason for selecting an interval of one half second was that this value corresponds approximately to one half period of the natural electro-mechanical oscillation of the average power system.

Most exciters are ordinary d-c, air-cooled generators, 250 volts, directly connected to the a-c generator and therefore operating at 3600 rpm. However, with the ever-increasing capacities of 3600-rpm generators and the corresponding increase in the size of the

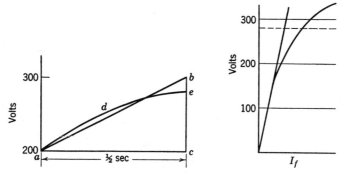

Figure 3-16. Curves illustrating the definition of exciter speed of response. Exciter build-up, curve ade. Area $abca$ = area $adeca$. Collector-ring voltage, 200 volts. Slope ab reaches 300 volts in ½ second = 50% increase in ½ second; therefore speed of response = 1.

exciters, the current-collecting parts of the exciters and generators are becoming difficult to build at these speeds. For instance, an 80,000-kw, 3600-rpm generator will require a 350-kw exciter, corresponding to 1400 amperes at 250 volts. A voltage of 357 volts has therefore been adopted for these large exciters to reduce the current. This value was chosen because it is the sum of the two standard voltages, 250 and 125 volts, and stations that have spare exciters at these voltages can use the two machines in series as a spare for the 375-volt exciter.

The mechanical difficulties of building large high-speed exciters have led to the use of exciters driven from the main unit through gearing. In some cases motor-driven exciters are used. Such units are usually equipped with a flywheel to ensure excitation for some time—a second or so—in case of loss of voltage to the motor. Some utilities have solved the problem by substituting static electronic excitation for the rotating machine. The electronic exciter consists of a power rectifier connected to an a-c source, usually the terminals of the generator to be excited. Its speed of response is extremely fast;

on the other hand, it is more complicated, more expensive, and probably less reliable than a rotating machine.

Parallel Operation

It is very seldom that a single generator feeds a utility load. There are usually several generators in each station and several stations operating in parallel. It is important to remember that if generators are alternating current, load cannot be transferred from one machine to another through merely changing their excitation, as can be done if generators are direct current. The output of the machines can be changed only through adjusting the turbine governors.

If the turbine governors are set to maintain constant speed irrespective of load, there will be no control over the division of load between turbine units. Conversely, with two or more generations on a bus, one cannot govern for constant frequency with variable load. Therefore, the governors of steam turbines are adjusted to have a drooping characteristic from no load to full load of the order of 3%. This is called the "slope" of the governor. It is independent of speed in the operating area so that it moves up or down parallel to itself in response to governor adjustment. The governor adjustment is done by a small reversible motor operated from the switchboard.

Every governor has a so-called dead band, a small band of speed within which the governor will not respond. It is necessary in order to give stability to speed control. With a fairly flat governor slope this dead band may correspond to quite a percentage of unit output, and it plays an important part in the problem of load and frequency control, which is discussed later.

Selection of Turbine Generator

Selection of a unit of the right size in an expanding utility is one of its most difficult problems. Assume that a utility estimates that it will require 60,000 kw additional power in 3 years' time and that the delivery time for the equipment in place is 3 years. The utility can buy a 60,000-kw unit with all necessary auxiliary equipment and by the time it is ready to go into operation presumably the additional 60,000-kw load will be there. However, it may be more economical to buy a 90,000-kw unit in the knowledge that the extra 30,000 kw will not be completely absorbed for 2 years. If this is a straightforward comparison the choice is relatively easy to make, the various propositions being reduced to their present worth, taking into account the cost of money, the idle investment, depreciation, etc.

The problem, however, is usually not so simple, being complicated by all sorts of factors that do not lend themselves to arithmetical solution. It may be advantageous, for instance, to postpone buying any unit, if some neighboring utility has excess power to sell for a year or two. It may happen that the State Utility Commission may consider the purchase of the larger unit as "overexpansion" and disallow the extra investment in setting up the rate structure. The state of the market is also a factor. In a falling market we would hardly buy a larger unit than necessary.

Efficiency does not enter largely into the problem as the steam rate for given steam conditions does not vary much with size in the sizes under consideration—60,000 kw and up. On the other hand, the choice of the steam conditions and heat balance depends largely on the cost of fuel. The labor required does not vary much with size, a factor which favors the larger unit. Likewise the overhaul time varies little with size, so that the out-of-service time for the larger unit is half what it would be for two smaller units.

All these factors must be considered in the selection of a unit. The practice frequently followed is to decide that the largest unit on a system should not exceed some arbitrary percentage of the system capacity, say 10%. It is advantageous to have several units of the same size on the system. As more units of the selected size are added, their individual capacity represents an ever-diminishing percentage of the system capacity. At some point in this reduction, the decision is made to go to a larger-sized unit.

CHAPTER 4

Generating-Station Auxiliaries

The purpose of the generating station is to convert the thermal energy of coal, oil, or gas into electricity. Conversion takes place in the boiler and turbine generator, but the reliability of the process is no better than the reliability of the many auxiliary pieces of equipment on which the continuity of operation depends.

The auxiliary equipment consists essentially of:

1. A low-voltage supply for auxiliary power.
2. Fuel-handling equipment.
3. Fans for combustion air.
4. Pumps for condensate and feedwater.
5. Pumps for condensing water.
6. Ash-handling and miscellaneous devices.

As discussion will show, the auxiliaries are of two kinds—the "essential," which cannot be shut down even for a short time without jeopardizing the continuity of service, and the "nonessential."

Auxiliary Power

The auxiliary power requirement in a modern high-pressure station amounts to 6 to 8% of the station output. The supply voltage is almost always 440 or 2400 volts, or both. Air-cooled transformers and oil-less switchgear are becoming standard practice. Since reliability of supply is of prime importance, it is usual to provide two sources, one "normal" and one "stand-by." In some cases only the essential auxiliaries are connected to the "normal" supply bus, and transfer relays are arranged to change over automatically from the normal to the stand-by supply in case of loss of voltage in the normal supply. Sometimes the two supplies are operated in parallel, and provision is made to eliminate one in case of failure in it. More frequently, however, the first scheme is adopted.

Generating-Station Auxiliaries

There are five principal methods of normal supply:

1. Transformers connected to the station bus.
2. Transformers connected to the generator leads.
3. One or more house turbine-generators (alternating or direct current).
4. Auxiliary generators coupled to the main generators—so-called shaft-end generators.
5. Supply from a neighboring station through transformers.

Transformers are the cheapest installation to supply auxiliary power in both initial and operating cost, but they are subject to system disturbances. If more than one bank of transformers is used, they should be connected to different sections of the station bus in order to minimize the effect of bus faults. House turbine generators are not affected by system and main bus disturbances but are considerably more expensive than transformers in initial and running cost. They exhaust either into the station heat cycle at some suitable back pressure or into their own condenser, depending on their capacity and the desired over-all station efficiency. Use of them has fallen off greatly in recent years. Shaft-end generators are used rather frequently with very large and important units, or where the station may become isolated from other stations on the system. They are more efficient than separate house turbine generators because they utilize the high thermal efficiency of the main turbine. They require no extra space, since the room they occupy is necessary to allow removal of the rotor of the main unit in case of repair.

Formerly boiler plants were definitely less reliable than turbine generators, and it was necessary to install more boilers than there were turbines. This required steam headers and a complicated array of valves so that any boiler could be shut down without shutting down a turbine. The remarkable improvement in boiler plants in recent years has made this practice unnecessary, and there has resulted what is known as the "unit type station," in which a single boiler with its auxiliaries supplies a single steam turbine generator. Quite frequently no steam valve is provided at all between the superheater and the turbine stop valve. The "unit" idea is frequently extended to include a transformer. The location of the station at some distance from the load center and the large capacity of the generators usually require that the main bus voltage be higher than the generator voltage. In this case the step-up transformer is three-phase and of the same capacity as the generator. There is no breaker or bus between the

66 Principles of Electric Utility Engineering

generator and transformer, and together they form in effect one high-voltage generator.

The elimination of a lot of steam piping and valves in this "unit-type station" and the simplification of the electrical connections result in important savings, and experience has shown that the over-all

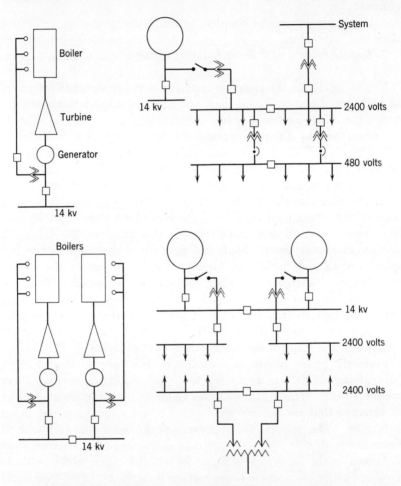

Figure 4-1. Typical auxiliary supply for unit-type stations.

reliability is quite satisfactory. The savings are reflected in the building costs where the boiler, turbine generator, and transformer can be laid out in a straight line with fuel and water supplied at one end (river or lake) and the transmission line taken out at the other end.

In stations of this type the essential auxiliaries are supplied through

Generating-Station Auxiliaries 67

a transformer connected to the main generator terminal leads as shown in Figure 4-1. A failure in the auxiliary transformer is equivalent to a fault in the generator leads and is taken care of by the differential protection around the generator.

In stations where a steam header is provided so that any combination of boilers can supply any steam turbine, it is not possible to connect the auxiliary supply transformer to the generator terminals if the unit idea is extended to include the auxiliaries. Figure 4-2 is a diagram

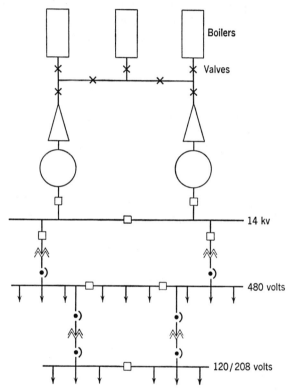

Figure 4-2. Typical auxiliary supply in multiboiler station.

commonly used in such cases. The transformers on individual sections are oversize and capable of carrying auxiliary loads on neighboring sections in emergency. Several operating procedures are possible for automatically switching from one section to another in emergency.

Where shaft-end generators are used, the nonessential auxiliaries are supplied from the auxiliary supply transformer and only the essential auxiliaries from the generator. This plan keeps down the size of the generator. The auxiliary supply transformer acts as stand-by for the auxiliary generator in case of emergency. With shaft-end generators the breakers are relayed so that the two sources of supply cannot be operated in parallel.

It is important to remember that a shaft-end generator cannot be operated in parallel with the main generator because its phase angle is fixed mechanically by the main unit.

In a few stations a secondary network is used to supply the auxiliaries. Although this system offers the maximum of reliability, its greater investment cost can be justified only by special local conditions, so that any widespread adoption is unlikely.

Fuel-Handling Equipment

The great majority of utility stations burning coal burn it in powdered form. In these stations the coal is fed from the coal pile on to a conveyor belt and carried up to a crusher. The crusher unloads it on to a second conveyor belt, which carries it up to coal hoppers in the station. From here the coal goes into the pulverizers, the quantity fed into the pulverizers being automatically weighed somewhere along the line.

The coal is reduced in the pulverizer to an extremely fine powder and blown into the furnace, where it is burnt in suspension. Heated air is supplied to the pulverizer under pressure and serves to remove moisture from the powdered coal (to prevent clogging) as well as transport the coal to the furnace. It is known as "primary air" and aids in combustion, but is of itself insufficient and therefore does not eliminate the forced-draft fan.

The amount of power required by a pulverizer varies widely with the kind of coal, fineness, and type of pulverizer. The range is from 5 to 35 kwhr per ton of coal. For a particular grade and condition of coal, the mill input versus tons per hour varies approximately as a straight line from no load to full load. The no-load input may be 15 to 40% of full-load input.

The pulverizing element of most types of pulverizers requires 200% torque for starting when full of coal, but, when the blower or exhauster is driven by the same motor, normal starting torque in terms of the combined horsepower may be sufficient. Squirrel-cage motors are most commonly used except for very large mills which may require wound-rotor motors.

Pulverizer mills are usually driven by open motors, although in some installations drip-proof or totally enclosed fan-cooled motors have been used because they are exposed to dripping water or excessive

Generating-Station Auxiliaries

coal dust. Wound-rotor motors should be equipped with enclosed collectors.

The crushers and conveyors also are driven by induction motors. The service is intermittent and may have to start under load. For this purpose, high torque (Class C) squirrel-cage motors are most frequently used except in the large sizes, where wound-rotor motors predominate.

NEMA recognizes four basic designs of squirrel-cage general-purpose motors defined as follows:

Class A with normal torque, normal starting current, and low slip.
Class B with normal torque, low starting current, and low slip.
Class C with high torque, low starting current, and low slip.
Class D with high torque, low starting current, and high slip.

With the progress made in motor design in recent years the characteristic curves of Class A and Class B have become practically identical at the same manufacturing cost. As a result Class A motors are seldom specified, and Class B motors have become the standard general-purpose motor of industry The starting torque of these motors varies from 120 to 180% of full-load torque, depending on size and speed; the pull-out torque from 200 to 250%; the starting current from 450 to 550% of full-load current; and the slip from 2 to 4% at full load.

Class C motors have two sets of rotor bars, an outer set with high resistance and an inner set with low resistance. On starting, when the rotor frequency is high and the reactance of the inner bars consequently also high, the current divides in the two sets of bars so that most of it flows in the outer high-resistance bars, thus providing high starting torque. As the machine comes up to speed and the rotor frequency falls, the current gradually takes the decreasing impedance path of the inner winding which results in a low running slip. The starting torque varies from 180 to 220% of full-load torque; the pull-out torque from 200 to 250%; the starting current from 450 to 550% of full-load current; and the slip from 3 to 6% at full load.

Class D motors come in two designs, one for intermittent and one for continuous duty. The intermittent rated motors are employed where the starting load is heavy and the starts are frequent, as in cranes, hoists, valves, elevators. The rotor resistance is high to give high starting torque with low starting current, and the slip is quite high, 8 to 13%. The continuous rated motor used with many types of machine tools has good starting characteristics, but the slip is more likely to be 5 to 8% at normal load.

Where the starting current must be kept low because of feeder or supply system limitations, or because of the large size of the motor, a wound-rotor motor is used. Introducing resistance in the rotor circuit makes possible the gradual application of a high starting torque without shock to the driven machine and with minimum

Figure 4-3. Totally enclosed tube-cooled motor. (1) Frame. (2) Cooling tubes. (3) Rotor. (4) Stator coils. (5) Fan and air intake. (6) Air outlet. (7) Terminal box. (Courtesy Westinghouse Electric Corporation.)

starting current. Since the slip for any given value of torque is proportional to rotor resistance, it is also possible to obtain speed control with a wound-rotor motor by means of an external rotor resistance, for example, liquid rheostat.

Practically all motors used for powerhouse auxiliaries have special insulation treatment resistant to dirt, oil, and moisture. Reliability is of prime importance in these auxiliary drives. Class B insulation (mica, asbestos, glass fiber, 130°C hot-spot temperature) is used in the larger motors, and where moisture or high ambient temperature is

Generating-Station Auxiliaries

particularly troublesome silicone insulation can be justified despite its high cost.

If the motors can be isolated from the coal dust or if frequent blowing out is assured, open motors can be used for indoor service. If blowing out is not possible, and in particularly bad locations, totally enclosed fan-cooled or force-ventilated motors should be used. Totally enclosed fan-cooled motors are now available from all major manufacturers up to 2500 hp, and their physical size is not unduly greater than that of corresponding open motors. Their cost may be double that of open motors. Even so a number of utilities have found that for essential auxiliaries of 250 hp or more in difficult locations, the greater reliability and much lower maintenance costs justify their use. Figure 4-3 shows a totally enclosed tube-cooled motor.

Stokers

The steaming capacity of most utility boilers has grown beyond the capabilities of mechanical stokers. They are, therefore, now used in only the smallest plants where the steaming rates of the individual boilers do not exceed 350,000 lb per hour. Stokers require a drive which can be varied over a wide speed range. It is also desirable to use a drive which will not vary greatly from the speed setting with change in load due to changes in the fuel bed. A four-to-one speed range is common, and six- or eight-to-one has been used.

The most common electrical drives for this purpose are d-c motors with field resistance or variable voltage control. Single or multi-speed squirrel-cage motors with mechanical or hydraulic variable speed transmissions are sometimes used.

Drip-proof or shrouded frame motors with special insulation or fan-cooled enclosed motors should be used because of the coal dust and moisture present.

Fans

Two major fans are required for the operation of the boiler—the forced draft fan which supplies the air required for combustion, and the induced draft fan which blows the products of combustion up the stack.

Roughly 21% of the air fed into the furnace is the oxygen necessary for combustion. The rest is nitrogen, which remains inert during combustion and passes uselessly up the stack. The quantity of air required for combustion is 8 to 12 lb per lb of powdered coal, depending on the heating value of the coal. This is the amount for which the forced draft fan must be built. This fan handles clean air, but the induced draft fan must handle gases at high temperature contaminated with ash and is therefore subjected to hard service and wear. Induced draft fans are designed specifically to reduce maintenance as much as possible. It is the hope of engineers finally to design the furnace of even the largest boilers so that it can be operated under pressure, and thus to eliminate the induced draft fan.

Large high-pressure boilers permit variations in load down to about half normal steaming capacity, so that the supply of fuel, air, and water must be adjustable to that extent. The adjustments are made automatically by the boiler control equipment. There are three methods of controlling the volume of air delivered by the fan:

1. Speed control.
2. Damper control.
3. Inlet vane control.

Speed control of fans is secured by use of adjustable-speed motors or of constant-speed motors and slip couplings. Damper control is the throttling of air flow in the fan outlet circuit. Vane control is a system wherein the air to the fan is controlled by adjustable vanes in the fan inlet. The vanes open and close, accelerating or retarding the twist given to the entering air. Inlet-vane control will govern fan delivery with an efficiency that compares favorably with any scheme of speed control if the range of speed control is not too extensive, and at a much lower cost. Figure 4-4 shows the motor input plotted against fan output for both damper control and vane control.

For a given fan size the capacity varies directly as the speed, the pressure as the square of the speed, and the horsepower as the cube of the speed. Where the resistance varies as the square of the volume as in most duct systems, the theoretical input will be proportional to the cube of the volume. A true comparison of the various control schemes, however, must be based on power input to the complete unit because the efficiency of most variable-speed motors decreases rapidly with speed. Table 4-1 gives a representative comparison of the power input for speed-controlled fans with different kinds of variable-speed drives as well as with damper and inlet vane control. The table is based on the assumption that all fans have the same efficiency at the point of maximum volume, and also that the pressure required varies as the square of the volume. As is to be expected a d-c drive with field and armature control comes closest to following the ideal cube law input.

In order to obtain the most economical drive for any given installa-

TABLE 4-1. TOTAL ELECTRICAL INPUT* VERSUS PER CENT MAXIMUM VOLUME FOR SEVERAL TYPES OF DRAFT FAN CONTROL

(Based on induced draft fan characteristics.)

% Volume	Single-Speed SC Motor with		8/10 Pole SC Motor with		SC Motor with Hydraulic Coupling	SC Motor with Magnetic Coupling	WR Motor with Liquid Rheostat	D-c Motor F + A Control
	Damper Control	Vane Control	Damper Control	Vane Control				
100	107.0	107.0	108.5	108.5	111.5	115.0	108.5	110.0
90	95.5	88.7	97.5	90.5	90.5	94.5	87.0	80.0
80	86.1	75.5	87.5	77.0	71.5	74.8	70.5	57.5
70	76.8	66.2	49.6	45.0	56.2	58.5	54.7	39.0
60	68.5	58.3	43.0	37.3	42.5	44.8	42.0	26.0
50	61.0	52.3	37.0	31.8	31.0	32.6	30.2	15.0

* Fan BHP at 100% Volume = 100.0.   SC = squirrel cage.   WR = wound rotor.   F + A = field and armature.

tion, it is necessary to consider the load cycle of the boiler, the total cost of equipment, and the effect of the scheme selected on the cost of the combustion control equipment. All the above schemes can be used for the control of both forced draft and induced draft fans. However, over-all economy and simplicity have led to a preference for constant-speed vane controlled forced draft fans, only about 30% of the stations built in recent years using variable speed. Of induced draft fans, about one third are vane control, one third damper control, and one third adjustable speed.

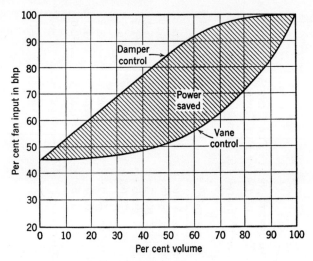

Figure 4-4. Induced draft fan performance with damper control and with vane control.

The starting torque requirement of fans is low. Motors with 60 or 70 per cent torque provide an ample factor of safety. The inertia of large fans may be sufficiently high to require specially designed rotors on squirrel-cage motors. The fan motors are not usually subject to splashing or dripping water, and any unnecessary enclosure is undesirable from a heating standpoint, but most operating companies feel that the extra protection afforded against falling objects by a box-type or shrouded frame more than offsets the slight increase in operating temperature.

Of those stations which have adopted variable speed the majority use adjustable-speed couplings. These couplings may be magnetic or hydraulic. In principle the magnetic coupling consists of a pole structure similar to a gear wheel with large flat teeth. The middle

Generating-Station Auxiliaries

of the gear wheel is cut out circumferentially to take a winding. With the winding excited, reiterative north and south poles are set up on either side of the winding. This member is mounted on the pump or fan shaft. On the motor shaft is mounted the second member, which consists of a drum surrounding the pole wheel with a small air gap. When the pole wheel is magnetized it will set up eddy currents in the drum, and if this is then rotated it will drag the first member around with it, the torque increasing with the intensity of the magnetic field.

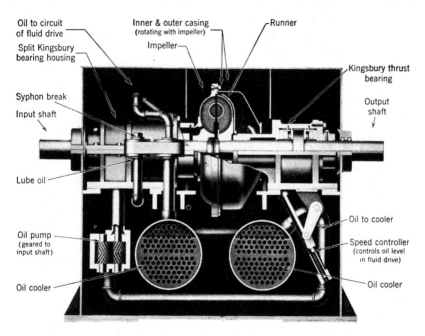

Figure 4-5. Principle of hydraulic adjustable-speed coupling. (Courtesy American Blower Company.)

The hydraulic slip coupling as built by the American Blower Corporation is shown in principle in Figure 4-5. It consists of an impeller and a runner in a single housing. The impeller rotates with the driving motor and acts as a pump. The runner receives the fluid and, acting as a reaction turbine, absorbs the kinetic energy developed by the impeller. Speed control is obtained through varying the quantity of fluid circulated through the impeller and runner by means of a small auxiliary pump. The coupling must have a slip of 2 or 3% in order to cause the fluid to circulate.

Pumps

Pumps constitute the greater part of the load on the auxiliary system. They may be divided into three groups: the boiler pumps, comprising the condensate and feedwater pumps with booster and recirculating pumps, if used; the circulating pumps for the condenser; a large number of miscellaneous pumps for feedwater make-up, ash sluicing, fire protection, drinking water, sumps, etc.

The simplest pumping system is one in which the pump operates against a constant head. The condenser which offers only frictional resistance to the flow of the circulating water is an example of such a constant-head application. In a system of this kind, centrifugal pumps driven by squirrel-cage motors are used almost exclusively.

In general, the starting of pumps does not present any difficulty since the starting torque required is only 15 to 20% of full-load torque, and the motors are usually designed to give 50 to 60% starting torque. The running parts are designed to stand full-voltage starting.

Many of the pumps are essential to the continuity of service. It is obvious that a flow of water must be maintained to the boiler if the flow of steam is to be maintained to the turbine. The flow of boiler feedwater depends on the condensate pump, the booster pump (if used), and the boiler feed pump. The boiler feed pump delivers water to the boiler at a pressure above the boiler pressure. A margin of 50% is allowed for low-pressure boilers (300 psi or less), tapering down to 25% for high-pressure boilers (1200 psi and up).

Most boiler feed pumps are multi-stage centrifugal pumps operating at 3600 rpm. The drive is by a direct-connected squirrel-cage motor. Where speed control is desired, a hydraulic coupling is used, or a wound-rotor induction motor with liquid rotor rheostat. It is desirable to operate wound-rotor motors at 1800 rpm rather than at 3600 rpm. Advantage is taken of this fact to build the pump for 5200 rpm. The combination of 1800-rpm motor, gearing, and 5200-rpm pump proves to be the most economical drive where speed control is used.

Final control over the flow of feedwater is obtained by a feedwater regulating valve which operates from both the flow of steam to the turbine and from the boiler drum water level.

The recirculating pumps used with forced circulation boilers are small compared with the feedwater pump and operate against a relatively small head. However, they must stand the full boiler pressure against atmosphere. They are constant-speed and throttle-controlled.

Generating-Station Auxiliaries

The steam condenser water box is invariably divided into two parts, each half with its own circulating pump. With half the condenser out of use, a fairly good vacuum can still be maintained,

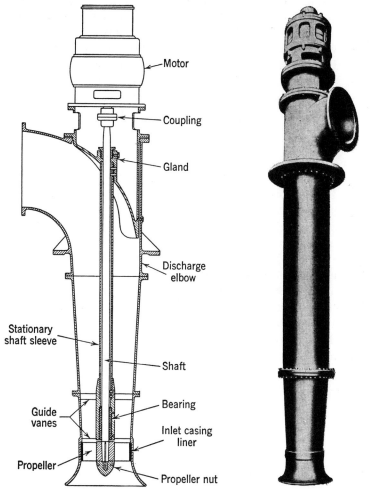

Figure 4-6. Vertical propeller pump for condenser-cooling water. (Courtesy Westinghouse Electric Corporation.)

especially in winter, and the turbine can carry full load. Despite this, the two condenser circulating water pumps are sometimes considered essential and a spare is provided. Vertical propeller pumps (Figure 4-6) are used rather than centrifugal pumps because for the relatively low heads required they permit much higher speeds. The motors are of the direct-connected squirrel-cage type.

The condensate pump is located close to the condenser below the water level in the hotwell. Where possible, it is a standard horizontal shaft centrifugal pump with squirrel-cage motor drive. If there is insufficient space between the hotwell water level and the station floor, a vertical shaft pit-type centrifugal pump may be used.

The nonessential pumps can be out of service for an indefinite period without impairing the service. For instance, the pumps in the feedwater make-up system may be out as long as there is water in the storage tank. The ash sluice pump can be down and the ashes can be removed by some other means. Sump pumps are important but do not require continuity of service.

Miscellaneous

House cranes use open wound-rotor motors. Compressors are usually driven by squirrel-cage motors. Valve operators use totally enclosed d-c motors which can be operated from the station battery under emergency conditions.

As mentioned above, both the power supply and the essential motor drives are important to the operation of the station, and more care is given to the problem of protection than to the same type of equipment used in industrial plants.

Shaft-end generators and supply transformers are protected with differential protection against internal faults and with inverse-time overload relays against external faults. The overload relays are cascaded down to the motors, so that as little equipment as possible is lost in case of failure. The differential protection will include as much as possible of the connecting leads to obtain protection against cable faults. Auxiliary generators will also be protected against loss of field.

CHAPTER 5

Hydroelectric Generating Stations

The great transmission systems of this country received their impetus as a result of hydroelectric developments. Forty years ago conditions favored such developments. Water-power plants costing $200 per kw or less were common at that time, whereas steam stations were relatively high in first cost, coal consumption per kilowatt hour was three times as much as today, and fuel oil was not readily available. As undeveloped water-power sites became economically less desirable, steam stations less in cost and higher in efficiency, and fuel oil and natural gas more generally available through pipe lines, steam stations rapidly outgrew hydroelectric stations in number and capacity. Today very few water-power sites can be developed at such low cost as to be competitive with steam stations in economic energy production. For this reason, almost all such developments of recent years have been undertaken by Government agencies. The agencies are in a position to include in the projects other considerations such as navigation, flood control, irrigation, and conservation of resources, which give them great social value.

As the water-power developments within easy reach of the load centers were utilized, and it became necessary to reach to greater distances for water power, only large developments could be considered, and stations of less than 100,000 kw became the exception rather than the rule, as witness Conowingo with 242,000 kw, Diablo with 135,000 kw, Fifteen Mile Falls with 140,000 kw, Osage with 200,000 kw, and many others. The developments of recent years undertaken by various Government agencies have reached gigantic proportions, for instance, Hoover Dam with 1 million and Grand Coulee with 2 million kw installed capacity.

Water-power developments are of two general types, those which dam the river and have spillway, control gates, and powerhouse in

80 Principles of Electric Utility Engineering

a more or less integrated unit, as in Figure 5-1; and those which short-circuit a part of the river fall by means of a pipe line, tunnel, or canal. In these latter developments the dam and spillway may be at some distance from the powerhouse, which is usually at the bottom end of a penstock bringing the water down from the pipe line, tunnel, or canal (Figure 5-2). In recent years many high-head plants

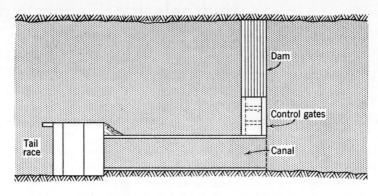

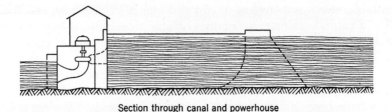

Section through canal and powerhouse

Figure 5-1. Small low-head water-power development.

have been located underground, not so much to protect them from bombing as for reasons of economy. Problems incident to the laying of penstocks, to building foundations, and to inclement weather are all made easier by underground location. Low-head plants invariably use the first scheme, damming the river; and high-head plants the second scheme, using penstocks. In between are medium-head plants which may use either scheme. In general terms heads 20 to 100 ft are referred to as low head; heads between 100 and 600 ft as medium head; and heads over 600 ft as high head. The lowest-head utility power stations in the United States are probably the McIndoes plant on the Connecticut River, with four 5000 hp units under 29-ft head and the Rock Island plant on the Columbia River with eight 21,000 hp units under 32-ft head. There are several plants in Cali-

Hydroelectric Generating Stations

fornia with units of 40,000 to 60,000 hp under 2200-ft head. Examples of intermediate head plants are Hoover Dam with 55,000 hp units under 480-ft head, Grand Coulee Dam with 165,000 hp units under 330-ft head, Pensacola Dam with 20,000 hp units under 115-ft head, Wilson Dam with 35,000 hp units under 92-ft head.

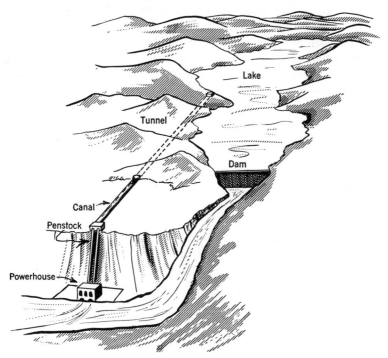

Figure 5-2. High-head water-power development.

Pump Storage Plants

Under certain conditions it is economical to use pump storage plants. Switzerland and Italy, where there is little or no fuel, have several such plants. In the United States there are two—the Rocky River plant of the Connecticut Light and Power Company and the Buchanan Dam plant of the Lower Colorado River Authority. In such plants at times of maximum river flow or during off-peak periods water is pumped into a lake on a neighboring mountain, or simply back from the tail race into the headwater to be used over again. The storage may be for lengthy periods to take care of seasonal shortages or for short periods to take care of daily peaks. In Italy the Alps have an excess of water in summer and a shortage in winter, whereas the Apennines have an excess of water in winter and a shortage in summer. The conditions therefore are ideal for seasonal storage and power exchange.

The Rocky River system has a single unit of 24,000 kw operating under a head of 230 ft. With the lake filled to capacity, it will give 24,000 kw for several hours a day for 6 months. The Buchanan Dam plant is a short-time storage system. It has three generators of 12,000 kw each and one pump of 13,500 hp. There is enough water to operate two 12,000 kw units at all times, but during dry seasons the third unit can be operated only part time. By pumping water back from the tail race to the headwater 10 hours a day, all three units can be operated 6 hours at full load, thus carrying peak load.

Automatic Stations

Another type of plant which proves economical under favorable conditions is the small unattended station which, except for a small maintenance cost, represents no expense outside the fixed charge on the capital investment, no regular labor being involved. The water wheel of such an installation has no governor, the gates being preset at some opening corresponding to the desired output of the generator. Suitable relays shut down the station in case of overload, overspeed, hot bearings, or any other kind of trouble.

In small stations induction generators lend themselves well to this kind of operation because of their simplicity. They require no exciter, voltage regulator, or synchronizing equipment. If any induction motor is driven at a speed corresponding to its slip *above* synchronism, it will generate power approximately equal to what it absorbs as a motor at the same slip *below* synchronism because the power transferred across the air gap is equal to the rotor copper loss divided by the slip, whether positive or negative. The disadvantage of induction generators is that they take their excitation from the system in the form of reactive current, thus lowering the system power factor. Since, however, they are always small in comparison to the system, the advantages outweigh this disadvantage.

Types of Water Turbines

The three types of turbines in general use today are (1) the reaction or Francis type of turbine, which utilizes the pressure of the water and the reactive force on the curved buckets, tending to change its direction; (2) the impulse or Pelton type of wheel, which utilizes the velocity and impact of a jet of water against buckets arranged around the rim of a wheel; and (3) the propeller or Kaplan type of

Hydroelectric Generating Stations 83

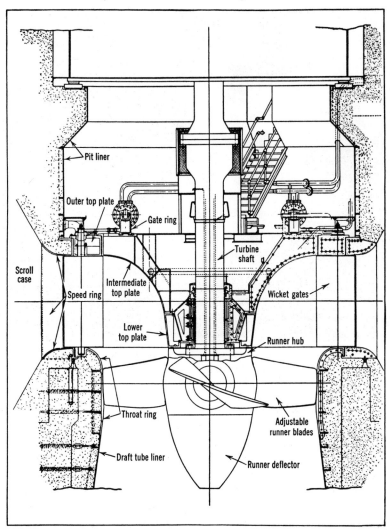

Figure 5-3. Cross-section through a Kaplan turbine showing speed ring, wicket gates, and runner. The opening of the wicket gates and the pitch of the runner blades are usually adjusted simultaneously for best efficiency at the prevailing head and load conditions. (Courtesy S. Morgan Smith Company.)

turbine, which is a reaction turbine of special design utilizing the axial flow principle. In general, Kaplan wheels are used for heads up to 100 ft, Francis wheels for heads between 70 and 1000 ft, and Pelton wheels for heads between 300 and 3000 ft. Figure 5-3 shows a section through a typical Kaplan or low-head turbine. The essen-

Figure 5-4. Illustration of a Francis turbine runner. The turbine casing and wicket gates are in principle the same as used for Kaplan turbines. (Courtesy S. Morgan Smith Company.)

tial parts are the scroll case, speed ring, wicket gates, and runner with adjustable blades. Figure 5-4 illustrates a Francis wheel used for medium heads. The casing and wicket gates are in principle the same as those used in the Kaplan turbine. Figure 5-5 shows the Pelton or impulse turbine used for high heads.

The design of a water wheel for any given conditions of head and output is based on an empirical value known as the "specific speed." This speed is defined as the rpm at which an identical model will operate under one foot head when reduced in size to develop one horsepower under one foot head. The rotational speed of the water wheel is obtained from the specific speed according to the formula

$$\text{rpm} = \frac{N_s \times H^{5/4}}{\sqrt{\text{hp}}}$$

where N_s = specific speed,
H = head in feet,
hp = horsepower of wheel,
$H^{5/4} = H \times \sqrt{\sqrt{H}}.$

Hydroelectric Generating Stations

The specific speed being a design factor, it varies among different manufacturers, but not widely. Figure 5-6 shows the band of specific speeds plotted against head for Francis and Kaplan wheels.

No curve is shown for Pelton wheels as the divergence of values is much greater than for Kaplan and Francis wheels depending on the number of nozzles used. The specific speed of Pelton wheels falls from about 12 to 4 in the range of heads from 300 to 3000 ft.

The flow of water is always given in cubic feet per second, Q, and the output of the turbine is therefore

$$\text{hp} = \frac{Q \times H \times \eta}{8.8}$$

where η is the turbine efficiency. Efficiencies of 90 to 92% appear to be easily attainable, and values up to 94% have been reported.

Figure 5-5. Shop assembly view of single jet impulse turbine with bypass. Most applications require long penstocks, and means must be provided to control surges in them on sudden changes in load. This control takes the form of deflecting the water from the buckets, or quickly opening a bypass nozzle which is then slowly reclosed in synchronism with the power nozzle. (Courtesy S. Morgan Smith Company.)

The limit of capacity for a given head is the size of the runner, which is preferably made in one piece. The scroll case is shipped in parts and assembled at the site. The largest one-piece Francis runner that can be handled on the railroads will give roughly:

200,000 hp at 190 rpm at 600-ft head,
175,000 hp at 129 rpm at 350-ft head,

100,000 hp at 100 rpm at 180-ft head,
70,000 hp at 97 rpm at 120-ft head,
40,000 hp at 95 rpm at 80-ft head.

The largest one-piece Kaplan wheel will give roughly

115,000 hp at 85 rpm at 100-ft head,
110,000 hp at 85 rpm at 80-ft head,
75,000 hp at 75 rpm at 55-ft head,
30,000 hp at 60 rpm at 30-ft head.

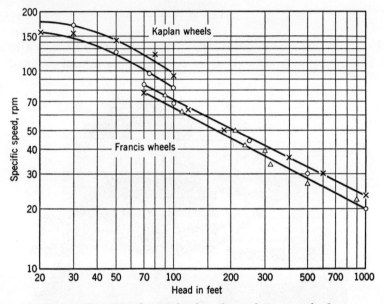

Figure 5-6. Specific speeds of Kaplan and Francis wheels.

The runaway speed of a water wheel, which is the speed the unit would attain with the gates wide open at no load, is about two times normal for Francis wheels and three times normal for Kaplan wheels. It is, however, controllable to some extent, and the generators are usually designed for 180 to 200% of normal speed.

The advantage of the Kaplan wheel, which resembles the propeller of a ship, is that it can be built with adjustable runner vanes. With change in load, the wicket gates and the pitch of the vanes are changed simultaneously so that a practically flat efficiency curve is obtained from 40% load to full load.

A further advantage is that at times when water is being stored and the water-wheel generators are being used to supply kilovars only,

Hydroelectric Generating Stations

the vanes can be turned to the fully closed position, at right angle to the axis, and the windage losses are reduced materially.

Governors

A water wheel, unlike a steam turbine, has a high inertia incompressible mass to counteract whenever there is a change in load. This presents problems not only in the design of the governor but

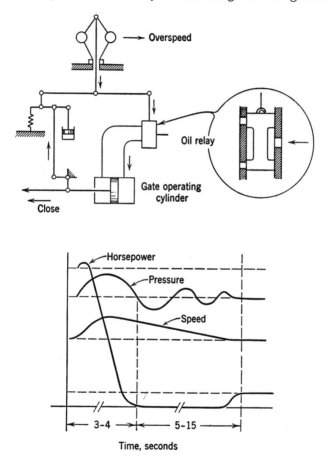

Figure 5-7. Principle of water-wheel governor control.

also in the design of the whole unit if stability is to be maintained during system disturbances. The governor is in the form of a servomechanism shown in principle in Figure 5-7. Any change in speed is reflected in the flyballs which through a lever move an oil relay to admit oil under pressure (150 to 300 psi) to one side or the other of a large piston in an operating cylinder. This piston moves to open or close the turbine gates, and in so doing, through a suitable linkage, to readjust the fulcrum end of the flyball lever to prevent hunting. Without this "inching" device the oil relay would swing up and down from its stable position and the unit would hunt.

When load is thrown off a water-wheel generator, a period of time must elapse for the unit to increase its speed to a value where the governor will operate the oil relay and build up pressure in the gate-operating cylinders to start closing the gates. This may take half a second. When the gates start to close, there will build up behind them a pressure which will momentarily increase the turbine output and further increase the speed. As a result of all this unavoidable delay, which is due to the incompressibility of water, on a drop in load the turbine may be increasing its speed for 3 or 4 seconds, and another 5 to 10 seconds will elapse before it is restored to normal. This is the reason why water-wheel generators have to be built with a high flywheel effect. This flywheel effect, WR^2, is specified by the water-wheel manufacturer chiefly on the basis of his experience. There is no way of calculating the required WR^2 from any theoretical considerations, the type of installation, size of system, penstock length, permissible pressure rise, and many other factors entering into the problem.

Generators

The generators used with water wheels are of two types: (1) the two-bearing design, in which the thrust bearing and one guide bearing are located above the rotor, with the second guide bearing below the rotor; and (2) the umbrella design, in which a single combination thrust and guide bearing is located below the rotor. The umbrella design, which is illustrated in Figure 5-8, has many advantages over the two-bearing design. It is somewhat less costly because the second bearing is eliminated. It requires less head room under the crane for dismantling and so reduces the cost of the building. The rotor may be unbolted from the shaft, which reduces the weight of the heaviest piece to be handled by the crane, leaving the shaft alignment undisturbed. The combined thrust and guide bearing is accessible from the pit. The bearing runs in a bath of oil so that all oil piping is eliminated. It lends itself readily to almost any scheme of ventilation. The air may be taken directly from above or below the rotor, or the machine may be totally enclosed and a recirculatory scheme used, with the coolers arranged conveniently around the periphery of the stator. If a scheme of excitation other

Hydroelectric Generating Stations

Figure 5-8. Umbrella-type water-wheel generator. (Courtesy Westinghouse Electric Corporation.)

than by means of a direct-connected exciter is used, the upper bracket may be omitted entirely. This may prove an advantage where it is desired to eliminate the building superstructure, and consideration should be given to motor-driven exciters where a dependable auxiliary source of power is available from which such exciters may be supplied.

Ventilation

All the large and medium-capacity water-wheel generators today are installed with a totally enclosed recirculating system of ventilation. The ventilating air is circulated through the generator, heat exchangers, and ducts by shaft-driven fans. The heat exchangers are cooled by water. The advantages of the recirculating system are:

1. Longer life of the insulation due to absence of dirt.
2. Reduction of fire hazard by the confining of the volume of air, which permits more effective use of fire extinguishing means.
3. Reduction of noise level.
4. Reduction of powerhouse temperature in the summertime.

Rotors

The rotor structure depends on the size of the generator and the required flywheel effect (WR^2). It may be a simple spoked wheel of cast steel, with the poles mounted on the rim and attached by bolts or dovetails. This design would be used in small machines running at low speeds. Where a large WR^2 is required in a small machine, the rotor may be made up of plates bolted together, with the poles dovetailed into the periphery.

In large generators the rotors consist of three parts: (1) the spider, which is usually of cast steel but may be built up of plates, (2) the rim of overlapping sheet steel laminations tightly bolted together, and shrunk on to the spider, and (3) the poles which are keyed in dovetails in the laminated rim. The rotor in umbrella-type machines is bolted to and centered on the shaft flange. The torque is transmitted by a large key or by dowel pins.

The thrust bearing, which has to carry the weight of the generator and water-wheel running parts and take up the water thrust, is either of the Kingsbury type or of the spring support type. The Kingsbury bearing consists of a collar attached to the shaft, and riding on stationary pivoted pads faced with babbitt. The whole bearing is immersed in oil, and a film of oil forms between the collar and the shoes and prevents wiping of the metal. In the spring-type bearing the collar rides on a steel disk with babbitted face. This disk is in turn carried on a large number of small springs, the whole being immersed in oil.

Damper Windings

The question is frequently raised whether or not damper windings should be used on water-wheel generators. In a polyphase generator

Hydroelectric Generating Stations

carrying a lagging power factor load and driven at constant speed, the field due to armature reaction will oppose the impressed field and rotate synchronously with it. The damper winding in the pole faces consequently rotate synchronously with this field, and the bars cut no flux and generate no voltage. Obviously they might just as well be left off. A properly designed water wheel is considered to produce a constant torque, that is, a torque that has the same magnitude at any instant of one revolution producing a constant speed. Its field structure will run at every instant in synchronism with the rotating field of armature reaction.

In contrast, a generator driven by a reciprocating engine has a torque which varies widely at different points of each revolution, depending on the piston position. Its field structure will therefore run part of the time faster and part of the time slower than the rotating field of armature reaction. In such a generator the damper bars cut the flux due to armature current, and currents flow in the damper which oppose the change of rotor position. In other words, the damper attempts to keep the generator running at an absolutely uniform speed despite a nonuniform torque.

This explains why damper windings are always used in generators driven by reciprocating engines and why they appear unnecessary in water-wheel generators. Nevertheless they are frequently provided in such machines for the reason that they do help somewhat during system disturbances. At such times the input into the generator from the water wheel is not matched by the generator output. Consequently electromechanical oscillations are set up which the damper winding by absorbing energy tends to slow down.

Damper windings become a necessity in remote-controlled and automatic stations to facilitate starting and synchronizing.

Brakes

To lift the rotor off the bearing, large generators are provided with combined hydraulic brakes and jacks. If the generator has been standing for any length of time, it is advisable to lift the thrust bearing collar off the pads to allow an oil film to form between them. A special control operated from the switchboard permits lifting and releasing the jacks rapidly when the machine is being started.

Auxiliaries

In general, the problem of providing auxiliaries for hydroelectric stations is not so complicated as that encountered in steam-station design. Those sections of the hydroelectric station devoted to the production of mechanical power are relatively simple; for this reason there are usually not so many normally running auxiliaries nor are there so many of the type from which an uninterrupted service is required to avoid a station shutdown caused by failure of the service alone. It is not, therefore, necessary to carry the duplication of units and provision for spare capacity to the same degree as is found in steam-generating stations.

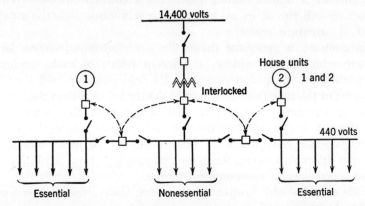

Figure 5-9. Typical diagram of station auxiliaries. *Essential:* Exciters. Governor oil pumps. Governor flyballs. Lubricating oil pumps. Fire pumps. *Nonessential:* Head gates. Crest gates. Sump pumps. Air compressors. Cranes. Intake gantry.

There are, however, problems peculiar to the design of hydroelectric auxiliary systems. The relatively isolated positions of many hydroelectric stations which feed their output over long transmission lines to distant loads call for a very careful consideration of the station auxiliary system layout. Important hydroelectric stations should be capable of independent starting and operation under all emergency conditions, especially during possible failures of ties to other stations. All the more important power plants have separate house units.

Selection of a station auxiliary one-line diagram is governed by the type of station, its capacity as compared with the total system capacity, its situation in the network in relation to the principal load, and its use as a base or peak load plant. They are relatively simple because of the few essential auxiliaries. A typical diagram is shown in Figure 5-9. Automatic transfer schemes for all auxiliaries may be provided in case of trouble in one bus section, or the essential auxiliaries may be put on stub buses and only those provided with transfer relays.

Hydroelectric Generating Stations

As in steam stations, the motors used are wherever possible polyphase induction motors of the squirrel-cage type. Except for cranes and hoists, variable speed is unnecessary. The practice of automatically restarting auxiliaries after a shutdown is fairly general, the automatic control taking the form of floats or pressure-type switches.

Combination of Steam and Water Power

It is a generally recognized fact that the value of an article is determined not by the cost of the article itself, but by the cost of the same article manufactured subsequently or by the cost of some equally satisfactory substitute. From the point of view of strict economics, no water-power site should be developed which, with its transmission, cannot deliver energy as cheaply as, or more cheaply than, a steam station located at the load center. As long as private capital was developing water power, this law of economics was strictly adhered to, and methods of determining the economic justification of such developments were carefully worked out. In recent years, however, the government has taken over the development of all major power sites and is going ahead with these developments without regard to the economics of the power aspect. The justification lies in the fact that the dams are necessary for flood control, irrigation, navigation, etc., and therefore, it is good business to spend the relatively small extra amount of money required to put in the power equipment and generate electricity. Only enough of the total capitalization is charged to power to make the cost of energy competitive with the equivalent cost of steam power at the load.

The characteristic of a water-power development is a high investment cost and a low operating cost. The characteristic of a steam station is a low investment cost and a high operating cost.

The cost of steam power per kilowatt-hour is

$$S = \frac{A}{8760f} + b$$

and the cost of water power per kilowatt-hour is

$$W = \frac{C}{8760f}$$

where A = fixed charges on investment + cost of operation and maintenance + cost of running auxiliaries and keeping plant hot, all as annual costs per kilowatt of station capacity,
b = fuel cost per kilowatt-hour,

Principles of Electric Utility Engineering

C = annual cost of water power per kilowatt delivered at the load center,

f = capacity factor.

An examination of these two formulae shows that since C, that is, the fixed charges on the dam, powerhouse, riparian rights, transmission line, etc., is much greater than A, water power will cost more than steam power at low values of f and less at high values of f, where the second term b begins to make itself felt.

The accurate determination of the economic justification of developing a water-power site in a utility territory supplied from steam stations involves much preliminary work by civil engineers to establish the history of the river flow—the variations from year to year, as well as the seasonal variations; the nature and value of the land to be flooded; foundations available for the dam; etc. Assuming that all this has been done, it is then necessary to strike an economic balance between steam and water power to give maximum economy. We might install enough generating capacity to take care of the maximum flow of the river during a short period. The incremental cost per kilowatt installed would be low, but the use made of the equipment (capacity factor) would also be low. Or we might install only enough capacity to use the minimum river flow. In this case, the incremental cost of the development per kilowatt installed would be high, but the capacity factor would also be high. Obviously between these two extremes lies an optimum value. The ratio of installed water-power capacity to the peak load of the system that gives the minimum annual cost of power supply is called the "economic hydro ratio."

In a hydroelectric development, transmission becomes a large factor of expense, and, in comparing such developments with equivalent steam stations at the load center, it is necessary to include the transmission as a charge against the hydroelectric plant. As will be shown later, the cost of transmission is substantially independent of voltage and amount of power transmitted, provided both are high. It is therefore possible to place the transmission costs on a mileage basis. This means that the further the water development is from the load center, the less desirable it becomes.

Quite frequently it becomes necessary to flood good farming country and populated areas in order to obtain a worth-while supply of power to justify any development at all. Care must be taken that the destruction of values in this manner does not exceed any alleged saving in the cost of power. Another point to remember is that in the

Hydroelectric Generating Stations 95

vast majority of industries the ratio of cost of power to total cost of finished product is less than 3%. The propaganda of public-power proponents must therefore be accepted with caution as regards "cheap water power" and its effect on the manufacturing economy of the country.

CHAPTER 6

Transmission of Energy

It is only in rare cases that a generating station can be located near a load center. A transmission system, therefore, becomes necessary even with steam stations. Transmission lines serve three purposes:

1. To transmit power from a water-power site to a market. These lines may be very long and, as hydroelectric sites become more remote, become still longer.
2. For bulk supply of power to load centers from outlying steam stations. These are likely to be relatively short.
3. For interconnection purposes, that is, for transfer of energy from one system to another in case of emergency or in response to diversity in system peaks. Such interconnections form the basis of the so-called integrated systems which are now a legal requirement of holding companies. Under a law passed some years ago, utility holding companies are permitted to control only such operating companies as are physically integrated in a given area.

The Federal Power Commission has set up a number of definitions applying to transmission, as follows:

Transmission means the transporting or conveying of electric energy in bulk to a convenient point, at which it is subdivided for delivery to the distribution system. Also used as a generic term to indicate the conveying of electric energy over any and all paths from source to point of supply.

Transmission system means an interconnected group of electric conductors and associated equipment for the transporting or conveying of electric energy in bulk to convenient points, at which it is subdivided for delivery to distribution systems.

Transmission line means a circuit in a transmission system.

Transmission-line capacity means the maximum continuous rating of a transmission line. The rating may be limited by thermal con-

Transmission of Energy

siderations, capacity of associated equipment, voltage regulation, system stability, or other consideration.

In the "uniform system of accounts" the FPC expands the foregoing definition of transmission somewhat to mention the items that may be included as part of the system in the accounting.

Frequency

The standard general-purpose frequency in North America is 60 cycles per second. Other frequencies used are 25 cycles for railway supply and around Niagara in the chemical industry. The Salt River Valley Water Users Association in Arizona has about 100,000 kw in 25-cycle generation. There is a small amount of 40-cycle generation in Maine, New York, and South Carolina; 30 cycles in Colorado, Michigan, and North Carolina; and some direct current in the older districts of a few large cities such as New York, Chicago, and Boston. Percentagewise, however, the whole capacity of these odd frequencies, except for the 25-cycle railway supply systems, amounts to very little, and wherever possible they are being eliminated.

In most foreign countries the standard frequency is 50 cycles. As a general-purpose distribution frequency, 60 cycles has an economic advantage over 50 cycles in that it permits without auxiliary devices a maximum speed of 3600 rpm as against 3000 rpm. Since the output of the generator is proportional to D^2LN, where D is the diameter and L the length of the active material and N is the speed, 60-cycle generators for a given output and number of poles can be made smaller in size because of the higher speed.

The principal gain, however, is in distribution transformers, of which a large number are used on any system. The induced voltage of transformers is proportional to the total flux linkage and the frequency. The higher the frequency, therefore, the smaller the cross-sectional area of the core. The smaller core also shortens the length of the coils and reduces the amount of copper used. In order to keep the core loss to a satisfactory value, however, the flux density used in 60-cycle transformers is usually lower than in 50-cycle transformers. As a result, the total saving in materials of 60-cycle transformers as compared with 50-cycle transformers will be not 20%, but rather 10 to 15%.

The only system under which any frequency lower than 50 or 60 cycles might be considered desirable would be in a long transmission line of, say, 300 to 600 miles. Such long transmissions have been discussed in connection with remote hydroelectric developments, and for

98   Principles of Electric Utility Engineering

these a frequency less than 60 cycles might be interesting because the inductive reactance of the line ($2\pi fL$) decreases and the capacitive reactance ($1/2\pi fC$) increases directly with the frequency f. This fact results in higher load limits, transmission efficiency, and better regulation. Full advantage of low frequency can be realized, however, only where the utilization is at low frequency. If the low-transmission frequency must be converted to 60 cycles for utilization, most of the advantage is lost because of the limitations of the terminal conversion equipment.

Direct-Current Transmission

The ultimate in reduction of frequency is d-c transmission, which has been given serious consideration and in special cases is still attractive. Widespread use, however, is unlikely because with the developments of recent years most of the objections to a-c transmission have been eliminated. The more apparent reasons that make d-c transmission attractive are the absence of the stability problem and the great saving in line copper as indicated by the simple relation:

$$P_{3\phi} = 3EI$$

$$P_{dc} = 2\sqrt{2} \times EI \times 1.06 = 3EI$$

where $P_{3\phi}$ is the three-phase power in watts,
P_{dc} is the d-c power in watts,
E is the voltage to neutral or ground,
I is the line current.

The factors $\sqrt{2}$ and 1.06 result from the fact that, with direct current, peak voltage and not rms voltage can be used and also that there is no skin effect. Clearly, therefore, with direct current the same power can be transmitted by two conductors as by three conductors with alternating current. Actually the gain is greater because d-c transmission would certainly involve high voltages, where corona becomes a determining factor. Corona appears on a conductor at some critical value of voltage. If the corona losses are assumed equal for alternating current and direct current, the limiting voltages in the two cases would be

$$\frac{\text{d-c line voltage}}{\text{a-c line voltage (rms)}} = \frac{2\sqrt{2}}{\sqrt{3}} = 1.63$$

Tests indicate, however, that a-c corona losses are greater than d-c losses; therefore, for a given conductor a higher d-c voltage can be used than indicated by the foregoing ratio.

Transmission of Energy

If cable is used, the difference in favor of direct current becomes more marked because with unidirectional dielectric stress it is possible to operate a cable to within 20% of its breakdown value continuously, since the losses due to ionization under d-c stress are negligible. Tests indicate that a 230-kv line-to-line a-c cable, for instance, could be operated safely at 400-kv d-c line-to-ground.

Direct-current transmission is not new. In 1906 a Swiss engineer, Thury, put in a d-c transmission from three power plants at Moutiers to the city of Lyons, a distance of 110 miles. The last 20 miles were in underground cable. The transmission was operated at a constant current of 150 amp, the voltage at full load being 100,000 volts. All the generators were connected in series and put in and out of circuit with variations in load. At the receiving end were a number of substations with all the motors connected in series. Each motor was connected to a three-phase generator, and distribution was in alternating current. The system operated for 30 years, being converted to alternating current in 1937. The cable after 30 years' use was said to be as good as new.

The d-c transmission that is discussed today is, of course, quite different. It is a constant-potential system with high-voltage rectifiers at the sending end and inverters at the receiving end. In the last 20 years many economic comparisons have been made between alternating current and the equivalent d-c transmission systems, but none of them indicated enough promise to justify the enormous development costs entailed in an actual project of any great magnitude. The main hindrance lies in the high cost and complication of the converters and inverters as compared with ordinary a-c transformers.

Industrial Frequencies

In many industrial applications, particularly in the machine-tool industry, 60 cycles does not permit a high enough speed, and 90, 120, and 360 cycles are frequently used. For finishing small holes, it has been found that belt-driven grinding and buffing tools do not provide a satisfactory finish. Therefore, frequencies higher than 60 cycles are necessary. Normal grinding and buffing surface speeds are 5000 and 7000 ft per minute. A one-inch hole, therefore, requires 19,000 rpm, a half-inch hole 38,000 rpm, and a ⅛-in. hole 152,000 rpm. Frequency is equal to $\frac{p}{2} \times \frac{\text{rpm}}{60}$, so that with a two-pole motor the small buffing tool would call for a frequency of 2533 cycles per second.

The Machine Tool Builders Association some years ago initiated with NEMA a program to standardize frequency. The standardization took the form of standard circuit voltages against speed and frequency so that any user can operate rotating equipment at the same speed with the same voltage from the generating equipment of any manufacturer.

For moderate output and relatively low frequencies, induction-type frequency converters are listed by manufacturers as standard equipment. They consist of two induction machines, both connected to the supply bus, but the second machine, with wound rotor and slip rings, is connected in reverse-phase rotation. This means that its rotor is driven in the opposite direction to its stator magnetizing field, and the frequency generated in the rotor will be the sum of the supply frequency and its frequency of rotation. For instance, assume that it is desired to supply power to a bank of ten machine tools of 3 hp each at 14,000 rpm from a 60-cycle source. The first machine (No. 1) of the frequency converter will be designed as a two-pole squirrel-cage motor and will drive the rotor of the wound-rotor machine (No. 2) at 3600 rpm. The stator of machine No. 2 is wound as a six-pole machine.

Machine No. 2 receives energy from two sources: 60 cycles of energy from the system which it imparts to the rotor as a transformer, and mechanical energy from machine No. 1. The frequency of this energy is

$$f = \frac{p \times \text{rpm}}{120} = \frac{6 \times 3600}{120} = 180 \text{ cycles}$$

so that the frequency at the slip rings is $180 + 60 = 240$ cycles. If then the machine tool motors are wound with two poles, their speed will be rpm $= \dfrac{120 \times 240}{2} = 14{,}400$. The relative outputs of the machines are proportional to the frequency of rotation supplied mechanically by machine No. 1 and the frequency of the system supplied to stator No. 2. Therefore, machine No. 1 gives to rotor No. 2

$$\frac{180}{240} \times 30 = 22.5 \text{ kw}$$

and machine No. 2 obtains from the system

$$\frac{60}{240} \times 30 = 7.5 \text{ kw}$$

Since these two values add in machine No. 2, it must be built for the total output; but machine No. 1 need be built for only 22.5 kw.

Transmission of Energy

In an actual design slip and losses would have to be taken into account.

Other industries use high frequencies. For instance, in the steel industry frequencies of the order of 960 cycles are used for steel melting and 9600 cycles for case hardening.

Choice of Voltage

The voltage selected for generation in power systems is almost invariably 13,200 to 14,400 volts. This is the most economical voltage for the great majority of generators built today. It is frequently the best voltage for city distribution, and, if the power station happens to be located within the city limits, its use materially reduces the number of substations.

With the largest generators built today—150,000 to 250,000 kva—a voltage of 14,400 presents difficulties in the design of the terminals. A current of 6000 amp appears to be about the maximum that can be handled with conventional terminals and in the circuit-interrupting devices. Voltages of 16, 18, and 22 kv are therefore used in these very large machines.

In the selection of transmission voltages, the principal consideration is not technical, but rather what is already in use in the general vicinity. The advantage of being able to make direct ties between adjacent systems outweighs a choice of voltage based on the lowest immediate cost. If, however, the contemplated transmission is remote from any existing system, the procedure of coming to a final decision involves several steps and a good deal of work. A rough calculation is made, using approximate formulae to obtain reasonably correct values of voltage and conductor size for the distance and power to be transmitted. Then a complete study of the initial and operating costs of the line is made for various assumed transmission voltages and various sizes of conductors.

Before any decision on voltage, reference should be made to the Report of the Joint EEI-NEMA Committee on Preferred Voltages to make sure that the value contemplated is a recognized voltage in common use. This report differs from preceding voltage standards in that it recognizes that there can be on a line no single value of voltage for all locations and hours of the day. Voltage varies, depending on both the distance from the point of supply and the load taken from the service. For this reason the report lists, in addition to the "nominal voltage," which is defined as "a nominal value assigned to a circuit or system of a given voltage class for the purpose of convenient designation," the "rated voltage," the "mode voltage," and "voltage spreads."

102 Principles of Electric Utility Engineering

The significance of this change will become apparent in the discussion of distribution circuits.

Kelvin's Law

In the lower voltages (up to 30 kv) for a given percentage energy loss in transmission, the cross-section and consequently the weight of the conductor required to transmit a block of power vary inversely as the square of the voltage. Thus, if the voltage is doubled, the weight of the conductor will be reduced to one fourth, with approximately a corresponding reduction in cost.

As the voltage goes higher, the saving becomes increasingly less because there are other considerations. Depending on the load power factor, the charging current may or may not increase the line current. The I^2R loss thus will be affected, and there may be leakage and corona losses if the voltage gets high enough. As an insurance against breakdown, important lines are sometimes built in duplicate. In such cases the cost of the second circuit must not be overlooked.

Lord Kelvin established a general principle regarding the most economical size of conductor, which has become known as Kelvin's Law. Properly applied, with all factors that are affected by conductor size included, the law can be quite useful in preliminary estimates. The law reads:

"The most economical area of conductor is that for which the annual cost of the energy loss is equal to the annual fixed charges on the capital investment of that part of the line which is proportional to the conductor area."

In the form of an equation, this law is

$$\frac{CI^2R}{1000} = pwa$$

where C = cost per kilowatt-year of energy wasted,
I = current,
R = resistance per mile of conductor,
p = cost per pound of conductor, including other costs which vary with conductor size,
w = pounds per mile of conductor,
a = per cent interest and depreciation.

In this equation the area of the conductor does not appear, but for the weight we can substitute the weight per circular mil (CM) per mile times the area (A) in circular mils. The weight per mile per CM for any given material is a constant. For a stranded copper

Transmission of Energy 103

conductor it is 16.3 lb per 1000 CM. Likewise for R we can substitute the area A divided into the resistance per mile per circular mil, which is also a constant for any given material. For copper it is 64,000 ohms.

Developing this idea, we obtain

$$\frac{16.3 \times A}{1000} \times pa = \frac{CI^2}{1000} \times \frac{64,000}{A}$$

from which

$$A^2 = \frac{4000 CI^2}{pa}$$

If the current I were a constant value flowing 8760 hours a year, we would obtain from this formula a straightforward answer. However, the current will vary with the load, and the load will vary according to some load curve (daily or annual). The peak value and the load factor correctly express the energy under the load curve, but the load curve is not a measure of the energy losses.

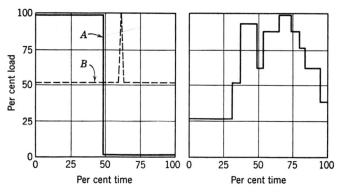

Figure 6-1. Three load curves with 50% load factor.

$$\text{Load factor} = \frac{\text{Average load over period}}{\text{Peak load over period}}$$

Consider the two extreme load curves A and B in Figure 6-1. A shows full load for 50% of the time and no-load for 50% of the time, whereas B shows half load for 100% of the time, with a very short-time peak. Both curves have a load factor of 50%, but the losses, I^2R, are obviously not the same for both. The two curves, however, can be used to draw up a practical "loss curve" for use when the peak load and the load factor are known. Consider again daily load curves of the type A and B, with a load of, say, 100 amp. Moving horizontally across the curves, we obtain for curve A losses equal to

$$100^2 R \times 6 \text{ hours}$$
$$100^2 R \times 12 \text{ hours}$$
$$100^2 R \times 18 \text{ hours}$$
$$100^2 R \times 24 \text{ hours}$$

which is an arithmetical progression. Moving vertically, we obtain for curve B losses equal to

$$25^2 R \times 24 \text{ hours}$$
$$50^2 R \times 24 \text{ hours}$$
$$75^2 R \times 24 \text{ hours}$$
$$100^2 R \times 24 \text{ hours}$$

which is a geometric progression.

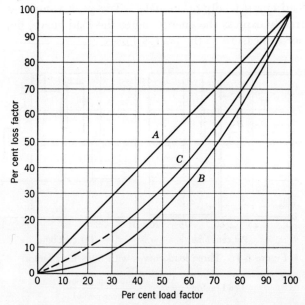

Figure 6-2. Loss factor to be used in loss calculations. *Example:* Peak load 20,000 kw, load factor 60%. For loss calculations use 43% or 8600 kw continuous load.

If these two curves are plotted against load factor as in Figure 6-2, we obtain loss curves which correspond to the extremes of load curves illustrated. The curve corresponding to all other load curves must lie in between them. For cable applications the National Electric

Transmission of Energy

Light Association adopted in the early thirties a curve C, based on the findings of a wide survey, the ordinates of which lie closely on a line one third the difference between the two extremes A and B above B.

If a maximum load of 10,000 kw and a load factor of 56% are assumed, the average load over 24 hours will be 5600 kw, but the losses as shown by curve C will be based on an average load of 4000 kw over 24 hours.

If the voltage is 34,500 volts, the value of energy 4 mills per kwhr or $35 per kilowatt-year, the cost of copper conductor 35 cents per lb, and the annual fixed charges 12%, then the area of the most economical conductor will be

$$A^2 = \frac{4000 \times 35 \times 67^2}{0.35 \times 0.12}$$

$$= 150 \times 10^8$$

$$A = 121{,}000 \text{ CM}$$

Using the nearest standard conductor, we would select a 2/0 conductor weighing 2170 lb per mile and with 0.481 ohm resistance per mile at 50°C. In making this selection, future requirements must be considered, as economy over the life of a line rather than at the time of installation is probably of greater importance.

Conductors

For overhead power transmission, the conductors used are stranded copper, hollow copper, ACSR (aluminum cable, steel-reinforced), ACSR with fillers, and hollow aluminum. Overhead distribution circuits are of stranded copper or aluminum. Copperweld, copperweld-copper, bronze, and steel are used on rural lines as current-carrying conductors, as overhead ground wires for transmission lines, for long river crossings, and for counterpoises.

For medium voltage, that is, subtransmission voltages, stranded copper has been universally used in the past, but the increasing cost of copper is creating a trend to aluminum. For high-voltage transmission, a minimum diameter of conductor becomes necessary to minimize corona effects. This proves an advantage for ASCR conductors, which for any given current-carrying capacity inherently have a large diameter as compared to copper. To meet this competition the copper conductor manufacturers brought out hollow conductors to obtain the large diameter without increasing the cross-sectional area, but here again the cost of copper is militating against its use.

Principles of Electric Utility Engineering

The electrical characteristics of these various conductors are published in table form by all manufacturers.

Corona

Corona on transmission lines comes about when the voltage on the conductor is sufficiently high to ionize the air surrounding it. The ionization depends on the voltage gradient (volts per centimeter) in the air surrounding the conductor and is therefore only indirectly affected by voltage. It may appear at 12,000 volts in a generator where the conductor is physically close to the grounded iron, but 100,000 volts may be necessary to make it appear on a transmission line. The ionized air surrounding the conductor is itself a high-resistance conductor. In effect, therefore, it increases the diameter of the conductor to the point where at its outside diameter the gradient is such that corona can no longer form.

The products of ionization of air are ozone and nitrogen-oxygen compounds. Corona of itself is not harmful, but in a generator where the conductor insulation contains organic materials (varnish and cellulose) and is in close proximity with iron, the oxidizing agents are harmful because they cause the organic materials to become brittle and seriously reduce their life. Moreover, the nitrogen-oxygen compounds combine with water to form acids that attack the insulation and metal. Mica is unaffected. Corona is no problem in hydrogen-cooled machines because ozone and nitrogen-oxygen compounds are absent, and ionized hydrogen is without effect on insulation.

In a generator, corona can appear in three places: in the slots between coil and iron and immediately beyond the iron; in the space between the coils in the end turns; and, if the insulation loosens up, in the voids thus created. The cure for corona in a machine is to paint that part of the coil in the slots and for a few inches beyond the slot with a high-resistance conducting paint and to make sure that the coil is in close contact with the iron. This latter is usually done by insertion of a mica strip in the slot between one side of the coil and the iron, forcing the other side of the coil against the iron.

On high-voltage transmission lines, the chemical effects of corona are usually not present, but corona is nevertheless objectionable because the leakage of electricity into the air (ionization process) represents a power loss, and because at sharp points and angles on the conductor and conductor hardware it emits short waves which cause radio interference. The voltage at which the air breaks down is called the "critical" or "disruptive" voltage. Various empirical formulae have been worked out to calculate this value. Such things

Transmission of Energy

as burrs, scratches, dirt, degree of smoothness, as well as diameter, humidity, and barometric pressure enter into the result so that accuracy is not to be looked for; but because of the importance of corona, a great deal of study has been given to the subject. Curves showing the minimum permissible conductor diameter plotted against operating voltage are found in several handbooks.

Conductor Spacing

The spacing of conductors in overhead lines depends on several factors: type and size of insulators, wind and icing conditions, and the economic weight given to performance against lightning surges. If maximum reliability is sought, the spacing loses its relation to the operating voltage, and then a medium-voltage line assumes most of the cost of a high-voltage line without the corresponding economy. In general, a compromise is adopted whereby the spacing is based on the dynamic voltage, with some allowance for reasonable performance against lightning surges.

TABLE 6-1. TYPICAL SPACING OF OVERHEAD CONDUCTORS

	Length, miles			Equivalent Spacing, feet			Number of Insulators		
	Ave.	Min.	Max.	Ave.	Min.	Max.	Ave.	Min.	Max.
14.4				3					
34.5				7	3	8			
69	35	25	100	11	5	18	5	4	8
115	40	25	100	17	11	25	7	6	11
138	40	25	140	19	12	20	10	8	12
230	133	45	260	31	17	35	15	14	20

Table 6-1 gives the "equivalent spacing" in feet and the number of insulators used on a large number of circuits in the United States, and may be considered representative of general practice. By equivalent spacing is understood the spacing that would give the same reactance and capacitance as if an equilateral triangular arrangement of conductors had been used. For design reasons it is usually impracticable to use a triangular arrangement. The equivalent spacing is obtained from the formula

$$D = \sqrt{ABC}$$

where A, B, and C are the actual distances between conductors. The formula would give rigorously correct equivalence only if the conductors were perfectly transposed. With transpositions normally

108 Principles of Electric Utility Engineering

encountered in practice, the error is of the order of 1 to 2%. With no transposition, the error may be nearer 10%.

Cables

In metropolitan centers distribution is usually by underground cables, and in recent years many underground cable transmission circuits have been put in where it was not economical to obtain the right-of-way for a high-voltage overhead line. Cables are in operation up to 230,000 volts.

The cable in most common use for power distribution is the paper-insulated, oil-impregnated type because of its low dielectric loss and low cost. Single-conductor cables are built up of layers of paper tape wound on a stranded conductor and impregnated with oil. Over this insulation is applied a tight-fitting extruded lead sheath. Three-conductor cables for power purposes may consist of three round conductors, each insulated with paper, the space between them being filled with a jute filler to round out the cable. Around this is more paper insulation, and over the whole a lead sheath. Where three-conductor cables have a conductor cross-section of more than 100,000 CM, the conductors are more likely to be sector-shaped. This design reduces the over-all diameter for a given cross-section of copper and provides better mechanical construction and proximity effect.

Proximity effect is the distortion of current distribution in the conductors due to the induction between the currents in the go and return conductors. It is directly proportional to the magnitude of the current and inversely proportional to the distance between conductors. Its effect is to increase the effective resistance of the conductors.

These three-conductor cables may be "belted" or "shielded." In the belted cable, the conductors are insulated with paper, cabled together with a filler piece, and then wrapped with more layers of impregnated paper (the so-called belt) before the lead sheath is applied. In the shielded cable, a thin metallic shielding tape is wrapped around the individual conductor insulation. Otherwise shielded is the same as belted cable. The purpose of the shielding tape on each insulated conductor is to control the electrostatic stress, eliminate corona formation, and improve thermal conductivity. Generally speaking, single-conductor paper-insulated cable is used up to 69 kv; three-conductor belted cable up to 15 kv; three-conductor shielded cable up to 46 kv. When the voltage exceeds 69 kv, special designs become necessary.

Other cable insulations are rubber, varnished cambric, and synthetic compounds. Rubber-insulated cables are commonly used for voltages up to 15 kv. The insulation thickness for a given voltage is greater than impregnated paper, but the cable is inherently more flexible and preferred for many applications. It has the disadvantage of aging more quickly if overheated and is easily damaged by oil. Special rubber compounds such as "Kerite" stand up better under difficult conditions.

Varnished cambric covered with a weatherproof braid is used in dry locations indoors for voltages up to 15 kv. It is somewhat more flexible than paper.

High-Voltage Cables

In recent years many high-voltage cable transmissions have been installed. At 69 kv and up, ordinary impregnated paper cable becomes quite unwieldy. Moreover, experience has shown that increasing the thickness of the paper is no warranty of better cables. One of the first ways of solving the problem, and one still popular, is the use of the Pirelli oil-filled paper-impregnated cable. In this cable the conductor strands are wound around an open helical spring so as to form a hollow conductor. Oil under pressure is kept on the cable by means of reservoirs so that no voids can form in the insulation. The oil is relatively thin, like transformer oil, and remains fluid at all temperatures. A cable of this type has been in service in Paris since 1936 at 220 kv. At these high voltages the cables are single-conductor. Three-conductor cables of the same general type with working pressures of 15 to 40 psi are manufactured for 34.5 and 48 kv.

Another cable which is widely used for high-voltage service is the Bennett pipe-type cable. In this type of installation three single-conductor paper-insulated cables are laid in a steel pipe 6 to 9 in. in diameter which is kept filled with oil under 200 psi pressure. Nitrogen gas is sometimes used instead of oil.

A third type of cable is the compression cable, which is the same as the Bennett type except that the three cables retain their sheaths, which may be lead or polyethylene, and the pressure is exerted on the sheath. This external pressure on the noncircular sheath acts on the paper insulation to prevent the formation of voids. The advantage claimed is that the threading of the cable into the pipe does not require such meticulous care.

Since the cost of these high-voltage cables installed will run from $75,000 to $200,000 a mile, they can be justified only where heavy loads are involved.

Cable Failure

Remarkable improvement has been made in the manufacture of cables, but cable failures are still fairly frequent. The study of insulation and insulation failure goes all the way back to Faraday and Maxwell, but even today a session on the nature of cable breakdown is good for a lively discussion.

The three theories of breakdown of solid insulation are:

1. Breakdown due to thermal instability (thermal breakdown).
2. Breakdown due to high electrical stress causing ionization by collision and slow disintegration of the material (ionic breakdown).
3. Breakdown due to mechanical disruption of molecules by the separation of positive and negative charges (disruptive breakdown).

Thermal breakdown is closely connected with the problems of dielectric losses. If these losses generate heat faster than it can be dissipated, the temperature will rise indefinitely until failure of some part occurs. It is thought that the breakdown of very hard materials such as glass and porcelain under voltage may be a thermal breakdown. There is a certain maximum voltage which a specimen will withstand indefinitely. For voltages above this value, breakdown must ultimately take place. The time of breakdown therefore, according to this theory, may be anything up to infinity.

The theory that best explains the breakdown of paper-insulated cables (except under impulse) is the ionic theory. Ionization begins in some part of the dielectric which in terms of its strength is more highly stressed than the remainder. This may be a little void which is not filled with impregnated paper or oil. Such a void is probably adjacent to the conductor. Failure comes about as a result of the progressive deterioration starting at one of these spots of initial ionization.

In an oil-filled cable the initial ionization may occur within the thin oil films between the impregnated tapes. This results in disintegration of some of the oil molecules with the freeing of gas, the formation of bubbles in which ionization wlil occur, and consequent failure.

Many investigators are of the opinion that the ionization type of breakdown does not hold for very short-time voltage applications such as impulse breakdowns. Test results do not show any relation to void formations. In very short-time voltage applications, the mechanism is better explained by an instantaneous physical rupture at some weak spot.

CHAPTER 7

Transmission Equipment

Stations transmitting power at voltages above that of the generator have a "switchyard" in which all the high-voltage equipment is located. It comprises the transformers, circuit breakers, isolating switches, lightning arresters, and buses. The equipment used in the substations at the other end of the transmission lines is essentially the same, but at the generating station there may be no breaker between the generator and the transformer (unit-type stations) whereas in the substations there are always breakers on the stepdown side of the transformers. The control of the breakers is in the station building at a switchboard which in addition to the control switches carries all the instruments and relays.

Transformers

The best indication of the economic importance of transformers is the fact that over 25 million kva in transformer capacity is added to power systems in the United States every year. There are two general designs of power transformers, the core type and the shell type, illustrated in Figure 7-1. In the core type of construction the steel punchings are arranged to form a single magnetic circuit consisting of "legs" over which the windings are mounted, and a top and bottom "yoke." In the shell type of construction, on the other hand, there is a single set of windings with two magnetic circuits encircling each side of the coils.

For voltages 138 kv and less, in capacities up to 10,000 kva, all power transformers are of the core type. For higher voltages and greater capacities they may be of the core type or the shell type, depending on the manufacturer. The core design in the medium sizes usually has coils in the form of circular discs. In the large high-voltage transformers the coils are more likely to be arranged in concentric layers around the low-voltage winding and core, with the line end at the outermost layer. The outer layer is surrounded by a shield

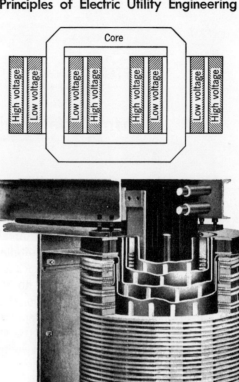

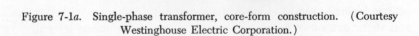

Figure 7-1a. Single-phase transformer, core-form construction. (Courtesy Westinghouse Electric Corporation.)

Transmission Equipment

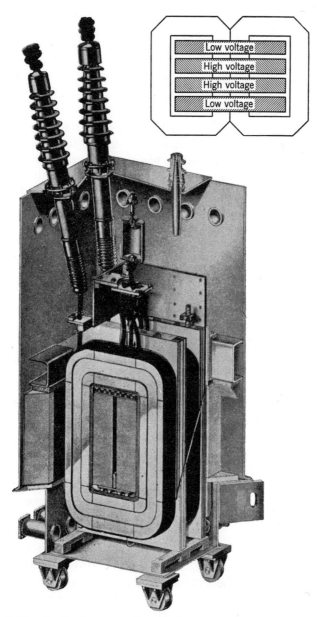

Figure 7-1b. Single-phase transformer, shell-form construction. (Courtesy Westinghouse Electric Corporation.)

connected to the line for the purpose of producing a uniform distribution of impulse voltage throughout the whole winding.

For electrical and physical reasons, the shell form lends itself well to large-capacity high-voltage applications. It inherently has a good surge-voltage distribution because the few large pancake coils give high coil-to-coil capacitance and low coil-to-ground capacitance. This fact results in good initial surge distribution, and the long natural period of oscillation assures quick attenuation. Since the insulation can be arranged in equipotential surfaces around the high-voltage coils, solid insulation must be punctured before failure can occur, as is indicated in Figure 7-2. It is in the physical arrangement, however, that the shell form shows a great advantage; since the iron surrounds the coils there is no necessity for clearance between the transformer proper and the tank, and the size and weight can be materially less than those of the corresponding core design.

If a sine wave of voltage is impressed on a transformer, the flux wave in it will also be of sine-wave form. However, the magnetizing current producing that flux cannot be a sine wave because of the nonlinear character of the saturation curve of steel. The magnetizing current contains in addition to the fundamental frequency wave harmonics of the 3rd, 5th, 7th, 9th order in decreasing amplitudes. In order to provide a closed circuit for these harmonics and prevent their circulating in the power system, it is common practice to connect three-phase banks and three-phase transformers with one winding (usually the low-voltage winding) in delta. Where the phase relation between two heretofore unconnected power systems calls for star-star transformers, a tertiary winding must be added to the transformers to accommodate the harmonics.

In recent years it has become common practice to use three-phase transformers rather than banks of three single-phase transformers. A three-phase unit is not only cheaper in first cost than three single-phase units, but is also more efficient, requires less floor space, and simplifies the required leads and structures.

For three-phase transformers to be operated in parallel, certain conditions must be met. They must generate equal voltages, and must have the same "impedance" to divide the load properly, the same phase rotation, and the same "polarity." The polarity depends on how the phases are connected. The ASA standard markings for the high-voltage terminals are H_1, H_2, H_3, and for the corresponding low-voltage terminals X_1, X_2, X_3, the phase rotation being H_1–H_2–H_3 and X_1–X_2–X_3. With these markings the standard polarity for delta-delta or star-star transformers is that which gives zero degrees dis-

Transmission Equipment 115

placement between the high- and low-voltage vectors of the reference phase, as indicated in Figure 7-3. For star-delta or delta-star transformers, standard polarity is that in which the high-voltage reference

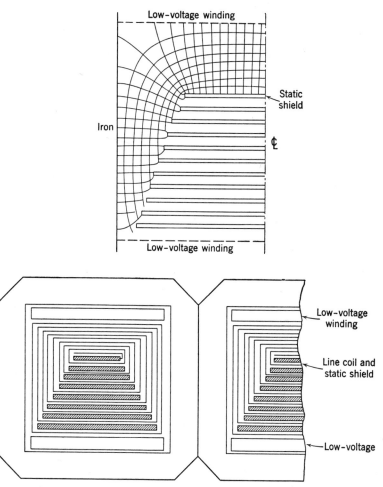

Figure 7-2. Schematic arrangement of insulating barriers in high-voltage shell-form transformer, resulting in electrostatic field and equipotential surfaces shown above.

phase is 30 degrees ahead of the low-voltage reference phase, irrespective of whether the transformer steps the voltage up or down.

The "impedance" of a transformer is not a simple impedance such as is found in, say, a piece of cable. When load is applied to a transformer, the current encounters an apparent impedance within the

116 Principles of Electric Utility Engineering

transformer which causes the voltage ratio to depart from the turns ratio. This apparent impedance is due to reactance derived from a leakage flux and a resistance due to all losses. It is expressed as a percentage of terminal voltage—primary or secondary line-to-neutral —or as a percentage of "normal ohms" where normal ohms is defined as "rated voltage, line-to-neutral, divided by rated current per phase."

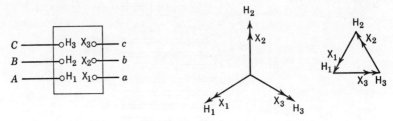

Figure 7-3a. Standard polarity for star-star and delta-delta transformers.

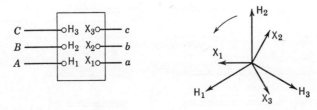

Figure 7-3b. Standard polarity for star-delta three-phase transformer.

The impedance varies widely with the voltage and class of transformer. From 69 to 230 kv it will range from 7 to 11% for self-cooled transformers and from 15 to 25% for forced-oil-cooled transformers.

The choice of transformer rating is in effect a choice of the type of cooling. National Electric Manufacturers Association a few years ago revised its designations of the basic types of cooling. In these designations O stands for oil, A for air, W for water, F for forced. The OA (oil to air) transformer is the oil-immersed self-cooled transformer in general use. With the addition of fans on the radiators, it becomes the OA/FA transformer. All the larger transformers (12,000 kva and up) are equipped with fans, and the extra capacity thus made available over the OA transformer is 33%. Many also have forced circulation of the oil and become OA/FA/FOA, in which case the gain in capacity will be 67% over the OA rating.

When a transformer is applied at a generating station in a unit arrangement, there is no likelihood of growth of load on it, so that the OA/FA rating should be matched with the maximum output of the generator. If it is a steam-driven unit, this would probably be the rating corresponding to 15 or 30 psi hydrogen pressure. Where the generating capacity is all bussed or when the transformer is applied at a substation, the probable growth of load must be estimated and a transformer must be selected of a size such that it will not have to be replaced or supplemented within the estimated period. The choice thus becomes largely a matter of judgment rather than an exact comparison of values.

A power transformer is inherently a device which will stand a considerable overload with relatively small loss in effective life, provided that its hot-spot temperature limit is not exceeded. For instance, during an emergency a transformer can be operated on most days of the year at 150% rated kva for 4 hours with less than 1% loss of life. provided that during the previous 20 hours the load has averaged not more than rated kva.

Noise in transformers arises principally from magnetostriction. Other causes such as poor joints, poor auxiliaries, etc., will not be present in well-constructed equipment. The new magnetic steels have less magnetostriction than the older steels, but they are capable of carrying one third more flux. Used at the same flux density as the older steels, they would be quieter, but to use the new and more costly steels thus, without taking advantage of their greater permeability, would not be economical. What this means, then, is that quieter transformers can be had at a price.

The National Electric Manufacturers Association has set down certain minimum noise levels which are attainable without undue increase in cost, but there is no guarantee that transformers meeting these levels will be acceptable in all circumstances. Much can be done to avoid complaints by care in the location of the transformers to avoid resonant surfaces, and by surrounding them with walls, trees, shrubbery, and other sound-absorbing means.

Switchgear

Switchgear is defined in the ASA Standards as "a general term covering switching and interrupting devices, also assemblies of those devices with control, metering, protective and regulating equipment with associated interconnections and supporting structures." The more important components of switchgear are circuit breakers and switches, fuses, control boards with instruments, meters, relays, and lightning arresters.

A switch is defined as a "device for making, breaking, or changing the connections in an electric circuit." In general, switches are rated

118 Principles of Electric Utility Engineering

in amperes and are capable of interrupting rated current at rated voltage. A fuse can be connected in series with the switch to take care of overcurrents.

An isolating switch is one intended for isolating an electric circuit from the source of power. It has no interrupting rating and is intended to be operated only after the circuit has been opened by some other means.

A contactor is a device operated other than by hand for repeatedly establishing and interrupting an electric power circuit. It also requires a fuse.

Circuit Breakers

A circuit breaker is a device for interrupting a circuit between separable contacts under normal or abnormal conditions. Ordinarily they are required to function only infrequently. They operate to close and open circuits under load when conditions are normal, and to interrupt circuits during abnormal conditions.

Over the years the time requirement placed on the circuit breaker has changed. It used to be that any breaker that successfully interrupted a circuit was accepted as satisfactory irrespective of the time it took to do so. As systems grew in extent and the problems of system stability became better understood, it was recognized that the speed of opening a faulted circuit was a major factor in increasing system stability. From then on, the efforts of the designers were devoted to shortening the time of interruption. In this they have been remarkably successful. In 1926 the time required to open the circuit was reduced from about 45 cycles to 25 cycles, in 1932 to 8 cycles, in 1940 to 5 cycles, and in 1954 it is 3 cycles. Time is the period from energizing of the trip coil to interruption of the arc. The ideal breaker would be one capable of extinguishing the arc at any voltage and current up to rating at the first current-zero after separation of the contacts.

The difficulty of opening a circuit is a function not only of the circuit voltage and the magnitude of the current, but also of the electrical constants of the circuit. It is more difficult to interrupt direct current, which must be forced to zero, than to interrupt alternating current, which has a natural current zero at each reversal in the direction of current flow. In order to extinguish a d-c arc, it is necessary to develop a voltage across the arc which will, for all currents down to zero, be greater than the voltage in the circuit. During the process of d-c interruption, all the magnetic energy stored in the inductance of the circuit must be dissipated as heat or transformed into electrostatic energy. If the attempt is made to interrupt a highly inductive circuit too rapidly by opening the contacts and lengthening the arc at very high speed, the voltage across the switch and inductance may reach dangerous values. The arc performs the function of dissipating the energy and controlling the rate of decrease of current.

The periodic nature of alternating current simplifies this type of interrupting problem, since in the process of reversing direction the current momentarily ceases to flow. At each current zero after the contacts have parted, the arc is extinguished and must be reignited. Since there is no stored magnetic energy in the self-inductance of the circuit at current zero, there is a theoretical possibility of opening the circuit at that moment without any arcing. In practice, however, the breaker may open at any phase position of the current, and an arc is drawn which is then interrupted at an early zero.

Dr. Joseph Slepian introduced the concept of extinction of the arc as a race at each current zero between the rising voltage and the recovery of dielectric strength of the arc path. He showed that in an oil breaker the arc and the gases created in the oil by the arc were necessary factors in interrupting the circuit. He described an oil breaker as essentially a gas-blast breaker and suggested that the way to improve oil breakers was to increase, not decrease, the rate of gas formation and to see to it that the newly formed gas was thoroughly mixed with the stream of ionized gas carrying the arc. This led to breaker designs in which the arc is forced to meet throughout its length ample quantities of fresh oil for gasification and arc-space deionization.

When an a-c breaker is opening a circuit at high power factor, the applied voltage and the current reach zero at the same instant, thus making the interruption easy by limiting the available restriking voltage. On the other hand, with low power factor, which is likely to be the condition during short circuit where the current is limited largely by inductive reactance, the applied voltage across the contacts at current zero is at its peak value. However, there is always some distributed capacitance in the circuit which must be charged up before full voltage can appear across the contacts, and hence a small interval of time is available for deionizing the arc space. The larger the capacitance to be charged up, the more slowly the voltage will build up. The rate at which it builds up is known as the "system voltage recovery rate" at that particular location, and it is dependent on the arrangement of the constants of the circuit. When the fault is removed from a simple circuit, the voltage will be accelerated toward normal, but will overshoot because of the energy stored in the system inductance. With no damping, the overshoot would reach twice normal voltage. Actually because of damping, normal voltage will appear after a few oscillations. If we assume a single oscillation, the

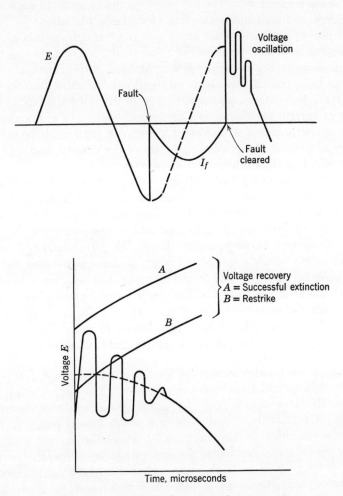

Figure 7-4. Principle of system voltage recovery.

recovery characteristic would be as shown in Figure 7-4, where A represents a successful extinction of the arc and B represents a restrike. In a practical system the transient recovery voltage is made up of a complicated combination of oscillations, and the calculation of such a transient is quite difficult. A typical recovery voltage curve of a transmission system is shown in Figure 7-5. In most circuit-breaker locations the rate of voltage recovery will not exceed 2000 volts per microsecond.

Opening of a faulted circuit is done by a breaker. Under certain conditions it can happen that the arc in the breaker will be interrupted but will restrike before the contacts have had time to separate far enough to prevent such a restrike. In other words, in the race between recovery of dielectric strength of the arc path and the rising voltage, the rising voltage has won. Restriking of the arc results in

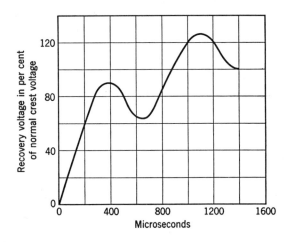

Figure 7-5. Recovery voltage curve of typical power system.

high transient voltages. Restriking or intermittent arcing can also occur in the fault itself and produce high transient voltages. If this restriking takes place in the switching device, the transient voltages are referred to as "switching surges." If it occurs in the fault, the phenomenon is called "arcing grounds."

A great deal of investigation has been carried out to determine the value of switching and arcing ground surges actually encountered in practice. It appears from this material that 10% of the surges encountered may reach four times system crest voltage.

In extensive systems with long transmission lines or high-voltage cables, it may become necessary to adopt strict operating procedures in order to avoid trouble from possible surge voltages. When this is done, the sequence of breaker operations is such that transformers are de-energized through enough cable capacitance to prevent dangerous voltages. A neutral connection to ground at the transformer will help reduce the duration and magnitude of transient voltages.

Types of Breakers

Breakers are produced in four general classes:

1. Low-voltage, 600 volts and lower, as used for station auxiliaries and at industrial installations.
2. Medium-voltage and medium-capacity, 4.16 to 13.8 kv, for distribution stations, small generating stations, and large industrials.
3. Heavy-capacity station breakers, 14.4 to 34.5 kv.
4. High-voltage outdoor breakers.

In the first class, for services below 600 volts, air breakers have been used rather generally. In recent years they have been much improved by the use of multiplate arc chutes to break up the arc and raise the interrupting ability. Carbon arcing contacts have been replaced by copper-tungsten alloy, and the main contacts are self-aligning silver-to-silver contacts instead of copper leaf, since oxidation of the silver, even if it takes place, does not cause excessive resistance and overheating. Improvements have also been made to the self-contained overcurrent and time-delay tripping mechanisms so that today selective tripping between breakers can be obtained. Standard ratings in these lines include manually and electrically operated breakers up to 6000 amp, although very much higher current ratings have been built for special services. Interrupting ratings go up to 80,000 amp rms.

The second class, breakers of medium voltage and capacity, comprises 4.16, 7.2 and 13.8 kv units with 25,000 to 500,000 kva interrupting ability. In this range oil breakers and magnetic blow-out air breakers are available, although the air breakers will not in all cases meet the AIEE basic insulation levels. The general construction of air breakers is somewhat similar to that of the low-voltage devices. As the breaker opens, auxiliary contacts insert a magnetic blow-out coil into the circuit, the field of which blows the arc into a multiple-plate arc chute. Contacts in general are silver-to-silver with copper-tungsten or silver-tungsten alloy arcing tips. In this range single-tank oil breakers with all three poles in one tank also are available. Where three tanks are used, they are usually supported by a common floor-mounted angle-iron framework.

In the third class oil breakers are being replaced by 14.4 to 34.5 kv air breakers for heavy-capacity station use, so that oil may be kept out of buildings to reduce fire hazard. These breakers are available in capacities up to 2.5 million kva. They are mostly of the air-blast type, the arc being literally blown into an arc chute. The

Transmission Equipment

opening and closing mechanism is pneumatic, and the movement of the contacts is interlocked with the blast valve to insure proper release of air. An air relay prevents operation of the mechanism unless adequate pressure is available.

In high-voltage service, the fourth class, the outdoor oil breaker is still by far the most widely used. The fire hazard is not so serious out of doors, and the interrupters have been so improved that the hazard has decreased rather than increased as the breakers have become larger in capacity. All the breakers built for this service by the various manufacturers meet the same specifications. Though the design of the interrupting elements differs considerably, the fundamental idea behind them remains the same, namely, to utilize the gas pressure created by the arc to force fresh oil into the arc path (with or without the aid of a piston) to extinguish the arc. Such an interrupting element is illustrated in Figure 7-6.

The principal manufacturers have built "oil-poor" or "low-oil-content" breakers. In these the outer housing has to be of porcelain, since the clearance to the live parts inside is relatively small. An outstanding example of this type of breaker is the General Electric Company's 287-kv breaker at Hoover Dam. It uses the "oil-blast" principle, the oil being forced through eight breaks in series mechanically by a piston actuated by the operating rod.

Compressed-air breakers have also been built for high-voltage service, chiefly to meet foreign competition. The high price of oil and metals abroad provides an inducement for the development of air breakers which is lacking here. They have a marked advantage over oil breakers in cleanliness of maintenance, and in not requiring auxiliary tanks in which to store oil during maintenance, but that is about the only advantage they have. They require large pieces of porcelain, possibly reinforced by other insulation, resulting in a structure not conducive to strength and ruggedness. Weatherproofing of joints is important. The operating mechanism is complicated by the necessity of isolator switches. Separate current and potential transformers are required, adding to the space requirements. No means has been found as yet of building them as cheaply as oil breakers. Whether or not, therefore, they will ever replace the highly developed dead-tank oil breaker is questionable.

Breakers are rated in accordance with Standard C-37 of the American Standards Association. Such a standard safeguards the purchaser in regard to the performance he is entitled to expect from the equipment, and it safeguards the manufacturer in that the limits of application are clearly defined. Breakers are given the following standard

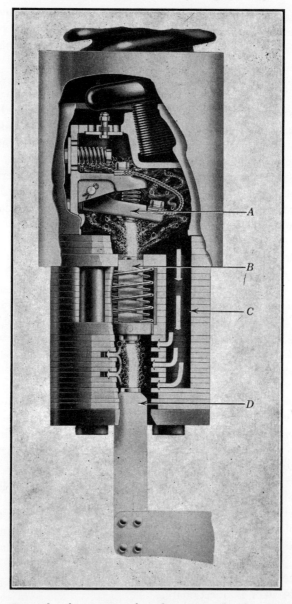

Figure 7-6. Principle of operation of modern circuit breaker interrupting element. When the breaker is opened the arc between contacts A and B creates gas pressure which forces oil through the channels C to extinguish the arc between contacts B and D, and thus open the circuit. (Courtesy Westinghouse Electric Corporation.)

ratings: voltage; continuous, four-second, and momentary current-carrying ability; insulation levels, low frequency and impulse; and interrupting ability, three-phase kva, maximum amperes and time in cycles.

Switchboards

The switchboards used by utilities are mostly self-supporting duplex boards, with front and rear vertical panels, both of which carry instruments and relays. Control desks or benchboards are frequently combined with such duplex switchboards. They are usually only control centers with the breakers and associated equipment mounted in more or less distant cells, which may be of concrete, brick, or metal.

The trend today is toward the use of metal-enclosed switchgear assembled and wired in the factory. It completely encloses the breakers, instrument transformers, buses, and connections, and means are provided for automatically disconnecting the live parts when it becomes necessary to gain access to the interior. This type of switchgear is built for indoor and outdoor service, that for the latter being weatherproof. In this form it is usually referred to as metal-clad switchgear.

Metal-enclosed switchgear is built in the low-voltage class for currents up to 6000 amp and in the voltage range 2300 to 15,000 volts, with interrupting capacities to 500,000 kva. This type of gear is used for powerhouse auxiliaries.

Cubicles are switchgear assemblies in which the breakers are permanently mounted and equipped with disconnects for isolating the breakers from the system. Their application is for heavy-duty breakers which are too large for use with simple disconnecting means.

The main conductors in a switchboard are known as buses. The size of the bus is determined by the current it must carry and the allowable temperature, which in switchgear is usually a rise of 30°C. Bus-bars are usually of copper ¼-in. thick and 3 to 6 in. deep. The bars are mounted vertically and spaced ¼-in. apart to form a single bus where one bar proves inadequate to carry the current.

The permissible maximum "hot-spot" temperature rise is 35°C above an ambient temperature of 40°C. When operating at this limit, a well-designed bus will have an average temperature rise over the surface of all bars of less than 30°C. Because of these temperature limitations, the current-carrying capacity of a bus does not increase in proportion to the number of bars. For instance, a single bar

3 × ¼ inches carries 1,000 amp. Four such bars spaced ¼ in. apart carries only 2200 amp. One 6 × ¼ in. bar will carry 1900 amp; four such bars, 4200 amp.

The bus insulators and supports must not only carry the weight of the copper, but also withstand the mechanical forces between them at times of short circuit, which can be very considerable. The force acting between two parallel bars, attractive if the currents flow in the same direction, repulsive if in opposite directions, in pounds per foot between supports, may be calculated from the formula

$$F = \frac{6 \times I^2}{d} \times 10^{-7} \text{ lb per ft}$$

where I = the short-circuit current,
d = the distance between the bars, center line to center line.

The formula gives results 50% higher than the theoretical value, but good practice dictates a large factor of safety since it can be obtained without much additional cost. It must be borne in mind that alternating currents cause eddy currents and hysteresis loss in metals near them. High currents, 3000 amp or more, usually require the use of nonmagnetic structural material, particularly when phase conductors are individually enclosed.

Control power to operate switchgear from the switchboard in generating stations and important substations is practically always direct current from a storage battery—usually 60 or 120 cells to give 125 or 250 volts. The larger batteries are charged by motor-generator sets, the small ones by rectifiers. To prevent the motor-generator set from reversing if the a-c power supply fails, a d-c reverse current relay is employed to open both the d-c and a-c breakers.

Supervisory control equipment is a type used very widely to provide remote control of substations or power plants over a few communication channels. One pair of telephone wires or one power line carrier channel is usually all that is required. The remote operator can perform such operations as closing or tripping circuit breakers, starting or stopping generators, adjusting the loading and voltage of generators, etc. By means of lamps, supervisory control provides the operator with continual indication of the situation in the controlled station, and with an alarm if automatic operation is interrupted. The performance of the many different operations requires the use of a coding system similar to that used in an automatic telephone system, except that certain refinements in the way of "check codes" are necessary to prevent getting a "wrong number."

Relays

A "relay" is defined in general terms as a mechanism which responds to some condition to cause the operation of some other device. There are relays on boilers responsive to air pressure, on turbines responsive to speed, on generators responsive to hydrogen density, and so on. To electrical engineers "relays" usually mean the complicated assortment of devices used to protect power systems against outage. Only the most widely used relays are described here. Their application is discussed in a later chapter.

The original purpose of relays was to protect apparatus. Nowadays this is probably a secondary consideration, the chief one being to insure continuity of service. Relays must be set for short-circuit protection and not for overload. Station operators must be relied upon to prevent overloads on equipment and transmission circuit.

Instrument Transformers

Relays receive energy for their operation from current and potential transformers connected to the circuit to be protected. The potential transformers reduce the system voltage to 120 volts, and the current transformers reduce the line current to 5 amp, two values for which relays can be readily built. Standard polarity of these transformers requires that when current is flowing toward the marked primary terminal, it is at the same instant flowing away from the marked secondary terminal.

In a current transformer the primary current is determined by the load current and not by the load on its secondary winding which is short-circuited by a low impedance meter or relay. The primary ampere-turns for excitation are therefore fixed by the current in the line. The core flux is small, and the secondary current changes proportionately with the primary except for the "ratio" and "phase-angle" errors due to the exciting currents. In wound-type current transformers with many primary turns these errors can be kept small, but in single-turn (through-type) transformers, the ratio and phase-angle errors can become considerable if the primary current is small. Where good accuracy is required as in metering, a current transformer should have 1000 ampere-turns or more. This stipulation means that the current should be at least 1000 amp where a through-type transformer is used. In relay work accuracy is not so important and through-type transformers are used down to 200 amp.

If the secondary of a current transformer is opened there will be no opposing secondary ampere-turns. All the primary ampere-turns

128 Principles of Electric Utility Engineering

will be used magnetizing the core, and high voltage will appear across the secondary winding. The secondary must therefore always be short-circuited before a meter or relay is disconnected.

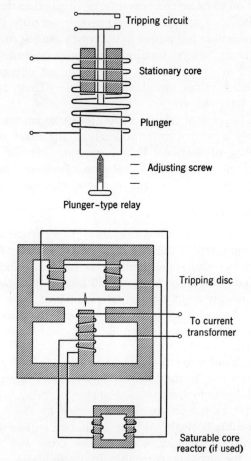

Figure 7-7. Principle of overcurrent relays.

Overcurrent Relays

The simplest relay and the oldest is the ordinary induction-type overcurrent relay, which is still the workhorse of protection schemes. The earliest designs consisted of an aluminum disc pivoted to run between the poles of an electromagnet of the shaded-pole type. The shaded pole principle is still used in some designs. In others a more complicated magnetic circuit is used for better control of the operating curve of time against current. The overcurrent relay (Figure 7-7)

Transmission Equipment

may operate with a time of tripping that is inversely proportional to magnitude of the current up to a certain point, beyond which the tripping time has a definite value irrespective of the magnitude of the current; or it may have a "very inverse" time-current curve, or a "definite minimum time" curve. Representative curves are shown in Figure 7-8. Tripping adjustment is provided for both current and time.

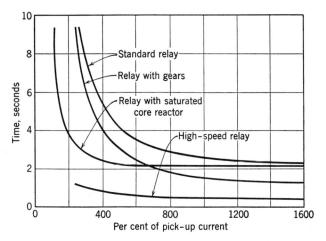

Figure 7-8. Representative time curves for various types of overcurrent relays.

These relays find a variety of uses for protection of both circuits and equipment. Where they are used for circuit protection one third to one half second difference in tripping time between successive relays should be allowed to obtain good selectivity.

An old form of relay still widely used is the plunger relay, in which a plunger floats in a solenoid. When the current exceeds a certain value the plunger is lifted and closes a contact. It is essentially an "instantaneous" relay. The value of current necessary to lift the plunger depends on its initial position in the coil. A common range of calibration is 1 to 4. For example, the relay will be set to trip at 5 to 20 amp.

Differential Relays

For removing faulty generators or transformers from the circuit, differential relays are universally used. The protective schemes are based on the principle of balancing the secondary currents in the current transformers at the terminals of the equipment so that under normal conditions, or during external faults, the current will circulate

130   Principles of Electric Utility Engineering

in the transformer secondaries. A protective relay connected in parallel with this balanced circuit will receive current only when a fault occurs in the equipment. Differential schemes of this kind are inherently selective and can be operated without intentional time delay.

Ordinary overcurrent relays can be used for differential protection schemes. However, the pick-up current value of the relay must be

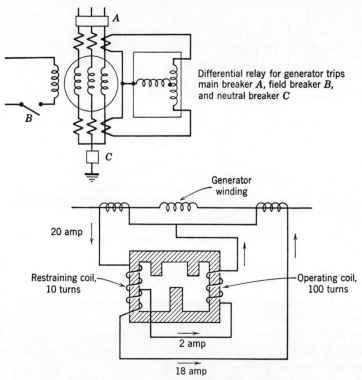

Figure 7-9. Principle of percentage differential relay.

set high enough so that any differential current due to unequal secondary loading of the current transformers will not cause the relays to operate. Unequal secondary loading may be due to unequal lengths of leads or unequal relay and meter loading. This differential current will be quite small under normal load, but it can become appreciable when short-circuit current is transmitted to an external fault. Hence, the relay pick-up must be set for this high value.

Where more sensitive protection is justified, as for all large machines, a so-called percentage differential relay is used. In this relay the current required to operate increases as the magnitude of the external fault current increases. For instance, if the relay is made to operate at 10% differential current, the difference in the secondary currents of the transformers must be greater than 10% of whatever value happens to be circulating through them. The principle of these relays is fairly simple. The magnetic circuit is similar to the one used in overcurrent relays but is energized by two sets of coils. A "difference coil" is energized by the differential current and produces a contact-closing torque. A "restraining coil" is energized by the through current of the two current transformers and produces a contact-opening torque on the disc. If the relay is made to operate on a differential current of 10%, the difference coil will have ten times as many turns as the sum coil, and the difference in the two currents must be more than 10% to produce operation. For example, in Figure 7-9 there is a torque equivalent to 20 amp and 10 turns tending to keep the contacts open, and one equivalent to 2 amp and 100 turns tending to close the contacts, so that the relay remains inoperative. If the current flowing through the operating winding is more than 10%, as it would be with a grounded generator coil, the torque due to the operating winding will overcome that due to the restraining coil, and the relay will operate.

These differential relays are also built to operate at very high speed. They can be applied to any kind of generator, delta- or star-connected, single- or double-winding, and to any kind of transformer, two-winding or three-winding, or to a combination of generator and transformer.

Directional Relays

In applications where the current can flow to a fault from two directions it becomes necessary to give the overcurrent relay a sense of direction if the loss of circuits is to be restricted to the faulty section. This is done by equipping the relay with a watt element which is responsive to direction of power flow. If the power flow is in the direction for which the watt element is connected, its contacts will close and permit the overcurrent element contacts to close. Otherwise the overcurrent contacts cannot close even though an overcurrent exists. The principle is shown in Figure 7-10.

Distance Relays

Most power systems today have extensive transmission networks with several sources of power. The relaying of such systems with simple overcurrent and directional relays even where possible would involve keeping a fault on a line for such long periods that the stability of the whole system would be jeopardized. It is imperative that faulted sections of transmission lines be switched out in the shortest possible time. Relatively complicated and costly relay schemes hence

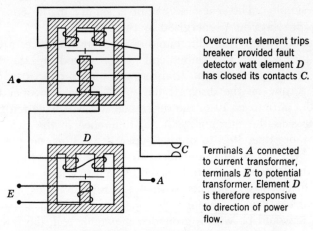

Overcurrent element trips breaker provided fault detector watt element D has closed its contacts C.

Terminals A connected to current transformer, terminals E to potential transformer. Element D is therefore responsive to direction of power flow.

Figure 7-10. Principle of directional overcurrent relay.

have been developed. Among the most widely used of these transmission line relays are the so-called distance relays—impedance and reactance types. These relays measure either the impedance or the reactance of the circuit between the relay and the point of fault as

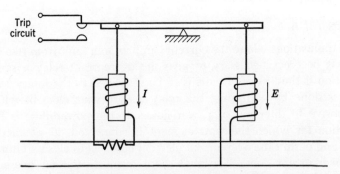

Figure 7-11. Principle of distance relay.

a measure of distance by balancing the voltage drop in the faulted circuit against the current of the faulted circuit. In principle the impedance relay is an overcurrent relay which is restrained by a voltage coil. The higher the voltage, that is, the greater the distance to the fault, the further the overcurrent disc has to travel to close the contacts. The reactance relay uses reactance as a measure of distance; it is more complicated as a quadrature voltage has to be established, but otherwise the principle is the same. It has the advantage that the voltage is unaffected by the fault resistance. In the impedance relay the aluminum disc tightens a spiral spring, the other end of which is attached to a lever pivoted at its middle point. One end of the lever carries the tripping contacts. The other end of the lever carries the core of the voltage-restraining solenoid. For any voltage E across the restraining coil the spring must be tightened to a definite point before the pull of the restraining coil is overcome. For a constant current I the time required to wind the spring and close the contacts will vary with the restraining voltage E. Therefore, the closing time is proportional to $E/I = Z$, or distance. The principle is shown in Figure 7-11.

CHAPTER 8

Power-System Fault Control

Utility systems are exposed to many hazards—lightning, windstorms, human errors, etc.—which may cause short circuits. The layout of the system and the design of the equipment must allow for these relatively infrequent abnormal conditions. If a system could be guaranteed against faults, its cost would probably not be one half of what it actually is. The removal of system faults and the limitation of fault current are what is referred to by the AIEE Protective Devices Committee as "fault control." The equipment primarily involved in fault control includes circuit breakers, reactors, lightning arresters, and relays. In the planning of a power system, fault control receives as much attention as the actual supplying of power to the load, for means must be worked out to continue supplying that power even though some part of the system may be in trouble.

The requirements of fault-control installations vary, of course, with the type of power system. They can be divided conveniently into three classes: large metropolitan systems, small metropolitan systems, and transmission systems.

In large metropolitan systems the power stations are likely to be within the metropolitan area, the distribution is at least partly at generator voltage, and the concentration of power is of necessity large. It becomes necessary to adopt ways and means of reducing as much as is reasonable the possible fault current and breaker-rupturing requirements. Reactance hence must be introduced between sources of energy, which can be done in several ways; for instance, by the use of reactors between generators and buses and between bus sections, by division of the bus into sections with breakers between sections normally open, by the use of double-winding generators.

In the smaller metropolitan systems some of the power stations are probably located outside the metropolitan area so that transformers intervene, and the problem of power flow and voltage regulation is likely to be more important than the magnitudes of the fault current and breaker duty. The breakers are of the less expensive types. It may be easier to avoid distribution over large areas directly from the generator buses.

In transmission systems the load centers are at some distance from points of generation, and there is inherently considerable reactance between generators by virtue of transmission lines and transformers. Even where the stations themselves are very large, the problem is made easier because it is possible to connect the generators directly to a transformer of equal capacity without any intervening breakers.

Supply Circuits

The method of connecting the supply sources and the feeder circuits to the bus is of major importance in fault control. In metropolitan systems where a large number of feeders leave the station at relatively low voltage two types of bus installation are in general use, the ring bus and the synchronizing bus schemes.

In the first scheme each generator supplies a bus section to which a number of feeder circuits—perhaps six or ten—are connected. The bus sections are connected together to form a ring, each being separated from its neighbor by a reactor. The scheme is shown in Figure 8-1. A more commonly used scheme is the synchronizing bus—sometimes called star bus. Here too each generator is connected to a bus section to which a number of feeder circuits are connected, but there is normally no direct tie between bus sections. Each section is connected to the synchronizing bus through a reactor, and this bus provides continuous supply to all feeders should a generator fail. Bus tie breakers, normally open, are provided between bus sections for use when one or more generators are out of service. The synchronizing bus also serves as a point where tie feeders from other stations can be connected and be available for symmetrical supply to all feeders through the reactors. Each feeder in both schemes has its own reactor between the bus and breaker. A reactor failure, now rare, is equivalent to a bus failure.

In metropolitan systems, under favorable conditions, the bus sections and stations can be tied together at the distribution network. If they are so linked, the different sources must not be out of phase by more than two or three degrees; otherwise the circulating current through the network will be objectionable. The scheme has the advantage of the usual low reactance in the main supply to the network, but there a high reactance results between groups of generators in case of a fault near a station bus.

In transmission layouts there are relatively few outgoing feeders, and the bus schemes are simpler. In a few "unit-type" stations the boiler, turbine generator, transformer, and transmission line are all

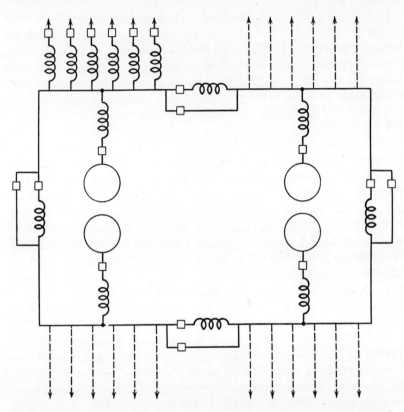

Figure 8-1. Ring bus for four-generator station. Disconnects not shown.

of the same capacity and operated with the bus-tie breaker open as shown in Figure 8-3a. In most cases, however, a duplicate bus is provided. Figure 8-3b shows a scheme that is frequently applicable and provides flexibility with a minimum of breakers.

High-voltage "breaker positions" from 115 to 230 kv cost $50,000 to $150,000, so that great care is taken in planning their application. Figure 8-4 shows a double-circuit transmission supplied from three generating stations. Scheme a is a fully sectionalized supply with sixteen circuit breakers, scheme b is a looped-in supply with twelve breakers, and scheme *c* is a bussed supply with nine breakers. Scheme *a* provides the maximum of flexibility and reliability, scheme *c* the

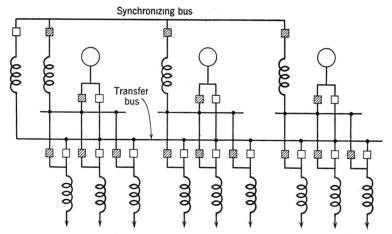

Figure 8-2. Synchronizing or star-bus scheme. The transfer bus is frequently omitted. Disconnects not shown.

minimum. With one line section out for maintenance in scheme *c*, a fault anywhere on the other circuit would result in the loss of the whole transmission.

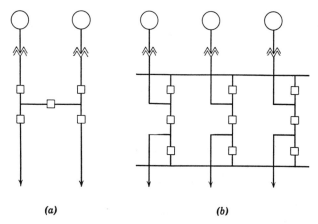

Figure 8-3. Unit scheme of bus connections applicable where generator, transformer, and line are of equal capacity. Disconnects not shown.

These typical bus arrangements shown can be multiplied into innumerable combinations to meet local conditions. It should be borne

138   Principles of Electric Utility Engineering

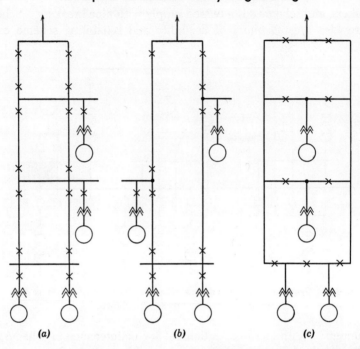

Figure 8-4. High-voltage bus schemes. (a) Fully sectionalized (16 breakers). (b) Looped in (12 breakers). (c) Bussed supply (9 breakers).

in mind, however, that complication to reduce the risk of outage frequently defeats its own purpose.

Application of Reactors

The reactors mentioned above are defined by ASA as "devices for introducing reactance into a circuit for purposes such as motor starting, paralleling transformers, and control of current." They are available in two forms, iron core and air core. Iron-core, oil-immersed reactors are used for higher voltage circuits, 34.5 kv and up, and where short-circuit currents are moderate, three times normal or less. Higher currents will saturate the iron in a reactor of reasonable size.

Air-core reactors, the type in more general use, are built for voltages up to 34.5 kv. In these lower-voltage circuits the short-circuit current is likely to be many times normal current. Air-core reactors have constant inductance. They take the form of flat bare copper cable coils securely held in concrete or fiber supports mounted on porcelain insulators. The problem of design is entirely mechanical because of the tremendous stresses the coils have to withstand under short circuit.

Power-System Fault Control

The rating of a reactor is expressed in kva absorbed at rated current and voltage. The reactance is expressed in "per cent" and is the ratio of the voltage drop across the reactor at rated current to the line-to-neutral voltage of the system. In a 20,000-kva, 14,400-volt circuit, a 6% reactor would be rated 400 kva, single-phase. The drop across it would be $0.06 \times 8300 = 500$ volts at 800 amp. The reactance would be $500/800 = 0.625$ ohm.

Care must be taken in locating reactors as they will induce current in neighboring metallic circuits. Cases have been reported of overheating of reinforcing rods in concrete walls due to reactors located close to the walls.

Application of Circuit Breakers

The application of circuit breakers requires careful study for economical as well as technical reasons. They are costly devices, especially in the higher voltage classes, and duplication must be avoided where possible. For a given set of system conditions, the fault current can be determined fairly accurately, but the difficulty comes in forecasting future circuit and generating-station changes which, during the expected life of the breaker, will materially change the duty it is expected to perform. One method used frequently to keep down breaker investment is to move breakers, as the system grows, from the heavier-duty locations near the powerhouses to the fringes where the interrupting requirements are lighter.

When a short circuit occurs on an a-c system, the current in the alternator rises to a maximum value within the first complete cycle. If the field remains unchanged, the current decays from the high initial value to a steady-state value symmetrical about the current zero line. This steady-state value is determined by the synchronous reactance of the generator. If the decay from the initial value is plotted on semi-logarithmic paper, it will be found to be made up of two exponential functions of time (two straight lines), one which lasts a very short time and is due to the subtransient reactance, and one which lasts appreciably longer and is due to the transient reactance of the machine. In other words, for the first couple of cycles after a short circuit the generator will behave as though it has a low reactance (subtransient x_d''). The reactance then assumes a higher value (transient reactance x_d') before settling down to its final value (synchronous reactance x_d) after all the transients are over. The time to reach final value varies widely with different types of generators.

Synchronous reactance includes armature leakage reactance and the reactance equivalent to armature reaction. The transient reactance x_d''

is that due to leakage reactance alone (with some correction due to induced currents in the field windings). The subtransient reactance x_d'' is a value which results from initial currents added to the transient component of current induced in the damper windings or similar circuits. In waterwheel generators with damper windings, x_d' will average around 37% and x_d'' 24%. Without dampers the two values will be almost equal, say, 35 and 32%. Turbo-generators with cylindrical field structures have the equivalent of damper windings, and good average values of transient and subtransient reactances will be (two-pole machines) $x_d' = 15\%$ and $x_d'' = 9\%$.

The reactance of transformers is expressed as a percentage voltage drop at rated current. The values of reactance of transformers will vary widely, depending on voltage and classification. In the OA class, 15 to 230 kv, the average reactance values may be 5 to 11%, but where transformer outputs are pushed by such devices as forced air and water cooling the reactance in large high-voltage units may reach 25 or 30%.

Reactance can be expressed in ohms or as a percentage drop of voltage at normal current and kva. For instance, in a single-phase circuit of 1000 volts and 20 amp one element of impedance may be 5 ohms or 10%, the latter meaning that with 20 amp flowing the voltage drop across the impedance will be 10% of 1000 volts. For conversion purposes the following formulae are convenient:

$$X = \frac{\text{kva} \times \text{ohms}}{\text{kv}^2 \times 10}$$

$$\text{ohms} = \frac{\text{kv}^2 \times X \times 10}{\text{kva}}$$

where kva = three-phase kva,
kv = line-to-line kv,
X = reactance in %,
ohms = to neutral.

In calculations care must be taken that all values of reactance are on a common voltage and kva basis; and in this regard the following simple relations are worth remembering.

Reactances in series are considered as having the same current flowing through them. For a given voltage this means the same kva. If, therefore, the reactances are reduced to the same kva, they may be added to give the total reactance. For instance, 4% on 10,000 kva, 5% on 15,000 kva, and 6% on 20,000 kva in series based on 30,000 kva give a total reactance of $12 + 10 + 9 = 31\%$.

Power-System Fault Control 141

Reactances in parallel are considered as having the same voltage drop through them. For instance, 4% on 10,000 kva, 5% on 15,000 kva, and 6% on 20,000 kva in parallel based on 6% drop give a total kva of 15,000 + 18,000 + 20,000 kva = 53,000 kva. Usually the calculation is based on the total kva of the system, or in this case on 10,000 + 15,000 + 20,000 kva. On this basis, the reactance is

$$\frac{6 \times 45,000}{53,000} = 5.1\%$$

Where a system involves many parallel and series reactances, it can usually be simplified by "star-delta" conversions to the point where only a single reactance remains between the point of supply and the point of short circuit. For instance, if in Figure 8-5 *abc* are three delta impedances and *xyz* the three equivalent star impedances, conversion from one to the other can be carried out by means of the relationship given in the figure.

Examples of Breaker Applications

The types of faults that can occur on a three-phase system are a short circuit across all three phases, a short across two phases, a short across two phases and to ground, and a short from one phase to ground. The last is by far the most common, but the application of circuit breakers is based on an assumed three-phase fault.

The procedure for calculating the highest value of short-circuit current at the breaker location is to use the subtransient reactance of the generators, the transient reactance of all synchronous motors, and the reactance of all other circuit elements intervening, for example, transformers and lines. Resistance is neglected, and the voltage used is the generator voltage to neutral expressed as 100%. With the voltage expressed in this manner the short-circuit current is a multiple of the rated or normal current. For example, if the total reactance between the source of generation and the fault is 25%, the short-circuit current will be 100/25 = 4 times normal.

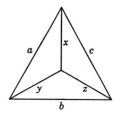

Figure 8-5. Delta-star and star-delta conversion.

Delta to star: $x = \dfrac{ac}{a+b+c}$

$y = \dfrac{ab}{a+b+c}$

$z = \dfrac{bc}{a+b+c}$

Star to delta: $a = \dfrac{xz+zy+yx}{z}$

$b = \dfrac{xz+zy+yx}{x}$

$c = \dfrac{xz+zy+yx}{y}$

142 Principles of Electric Utility Engineering

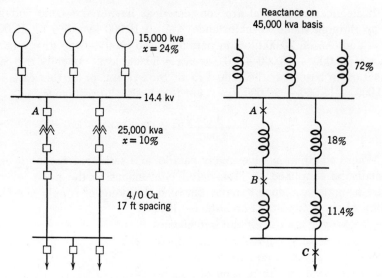

Figure 8-6. Example of determination of short-circuit current. System diagram at left; reactance diagram at right.

Assume the case of three waterwheel generators of 15,000 kva each, 14,400 volts, supplying a substation 40 miles away over two circuits of 115 kv as shown in Figure 8-6. The generators have 24% subtransient reactance, the two three-phase 25,000 kva transformers have 10% reactance, and the lines are 4/0 copper with 17-ft spacing. The interrupting duties are required at locations A, B, and C. The station capacity is 45,000 kva, and this will be used as the kva base. Normal current is then 1800 amp at the generator bus and 225 amp at the high-voltage bus.

The reactance of the transformers on 45,000 kva base is

$$x = \frac{45,000}{25,000} \times 10 = 18\%$$

The reactance of the lines is obtained as follows:

4/0 copper conductors at 1-ft radius x_a = 0.497 ohm per mile
for 17-ft spacing x_d = 0.344 ohm per mile
total x = 0.841 ohm per mile

or, for 40 miles, 33.6 ohms.

$$x = \frac{\text{kva} \times \text{ohms}}{\text{kv}^2 \times 10}$$

$$= \frac{45,000 \times 33.6}{115^2 \times 10} = 11.4\%$$

Power-System Fault Control 143

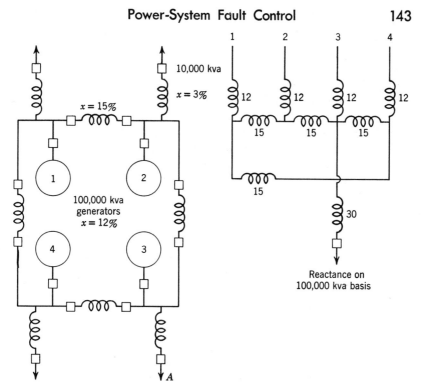

Figure 8-7. Determination of short-circuit current in station with ring bus. Further development in Figure 8-8.

With these values it is possible to convert the system diagram at the left of Figure 8-6 into the reactance diagram shown at the right, and the three-phase fault current at A is equal to

$$I_A = \frac{100}{24} \times 1800 = 7500 \text{ amp}$$

At B the three generators are in parallel as also are the two transformers. The reactance to that point is therefore $24 + 9 = 33\%$, and

$$I_B = \frac{100}{33} \times 225 = 675 \text{ amp}$$

At C the reactance of the two lines in parallel must be added, and the total reactance becomes $24 + 9 + 5.7 = 38.7\%$. The fault current at C is, therefore,

$$I_C = \frac{100}{38.7} \times 225 = 582 \text{ amp}$$

As another example consider the ring bus shown in Figure 8-7, with four generators of 100,000 kva each, 14,400 volts, 12% subtransient

144 Principles of Electric Utility Engineering

reactance. Each bus section has ten 10,000-kva feeders, in each of which is a reactor rated at 3% on 10,000 kva. Between any two bus sections is a reactor rated at 15% on 100,000 kva. Assume a three-phase short circuit on a feeder at A.

As a convenient base 100,000 kva will be used, although any base can be used, provided that one is consistent. The reactance diagrams take the form shown in Figure 8-8, and it will be seen that the method of substitutions must be employed to obtain the over-all equivalent reactance from the sources of supply to the fault at A. The consecutive steps are shown in Figure 8-8a to h, and the final reactance value is 36%. The normal current with 100,000 kva and 14.4 kv is 4000 amp, and the short-circuit current is therefore

$$I_A = \frac{100}{36} \times 4000 = 11,100 \text{ amp}$$

It will be noticed that the current is limited almost entirely by the feeder reactor which has 30% reactance on 100,000 kva base. Without these feeder reactors the short-circuit current would be

$$I_A = \frac{100}{6} \times 4000 = 67,000 \text{ amp}$$

This is equivalent to a three-phase short circuit on the bus, with generator No. 3 feeding directly into it, and the other generators feeding into it through the bus reactors in parallel.

The normal operating current and three-phase short-circuit current as determined in the foregoing examples are used as the basis for selecting circuit breakers. ASA Standard C37.6 gives the voltage, insulation level, current, and interrupting ratings of circuit breakers. In selecting the breakers, however, the values as determined in the examples must be multiplied by an appropriate factor, depending on the speed of breaker opening and the location of the breaker. This multiplier is as follows:

Speed of Breaker Cycles	Fault at Generator Voltage	Fault beyond Transformer
8	1.1	1.0
5	1.2	1.1
3	1.3	1.2
2	1.5	1.4

It would of course be just as correct to express the short circuit in kva as in amperes. In the last example, for instance,

$$\text{kva}_A = \frac{100}{36} \times 100,000 = 278,000 \text{ kva}$$

Power-System Fault Control 145

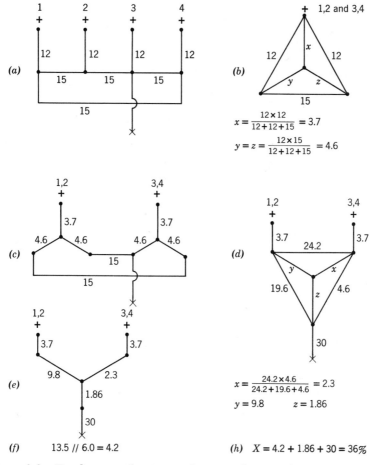

Figure 8-8. Development of reactance diagrams of system shown in Figure 8-7.

The ASA standard gives breaker interrupting ratings in kva and in amperes. The limit in breaker rating, however, is current, not kva. With decreasing voltage we may increase the current interrupting value but only up to a certain maximum permissible value.

Phase Segregation

As shown by the second example, above, the short-circuit energy in a metropolitan station can become enormous if nothing is done to limit it. In this example, for instance, a bus short circuit would be

$$\frac{100}{12} \times 400{,}000 = 3{,}330{,}000 \text{ kva}$$

146   Principles of Electric Utility Engineering

Generators are sold with a guarantee that when new they will withstand a three-phase short circuit at their terminals with full field, provided that the duration of the short circuit is not such as to damage the windings thermally. In other words, the mechanical parts and the winding bracings will stand a three-phase short circuit. A method used for many years, and still used, to reduce the short-circuit current to what the breakers and machines can handle is to segregate the phases so that only a short circuit to ground can occur, and then by means of a neutral impedance to limit the fault current to a value not greater than the three-phase fault current of a single generator.

In the early days this phase segregation was carried out by locating the leads, switches, and breakers of the three phases on three separate floors or in separate vertical galleries of the switchhouse, so that contact between two phases was not possible. Many switchhouses of this sort are still in operation, but no more are being built, for the improvements in switchgear have been such that the much greater expense of these lay-outs cannot be justified. Where phase segregation is used today it consists in running each phase circuit in a grounded metallic tube of nonmagnetic material, so that if a conductor fails it can fail only to the grounded tube.

Double-Winding Generators

A generator wound with two sets of windings brought out to separate terminals is a "double-winding generator," which in effect amounts to two generators in one frame excited from a common field. There are several ways in which windings can be paralleled in such machines, but some of them are not adapted to two-circuit operation because of magnetic unbalance, or overheating, or other difficulties if the two circuits become unequally loaded. In order to bring the terminal voltages of the parallel circuits into phase, certain slot combinations are necessary, and to obtain these may require somewhat impaired performance and increased cost. Therefore, double-winding generators are not always a desirable solution to the problem of reducing breaker current and interrupting duty. Nevertheless, the scheme is used rather extensively as 6000 amp appears to be the limit of current that can be handled through conventional terminals in a hydrogen-cooled generator. The reduction of the normal operating current per terminal is more likely to be the reason for adopting double-winding generators than is any consideration of interrupting duty. In a properly designed generator there can be a great difference in the relative loads on the two halves of the winding without inconvenience. Such a machine can in fact be operated with only

Power-System Fault Control

one winding in service if the other winding is in good condition. With a fault in one winding the whole generator must be shut down.

As regards short-circuit duty there is some advantage in a two-winding machine. Any rotating machine, synchronous or induction, can be treated as a transformer, with the frequency of one winding reduced to zero or near zero by rotation. For all transient phenomena, the machine is a transformer with primary and secondary, each having its own resistance and reactance. Where the armature winding is divided into two separate circuits, the equivalent circuit therefore becomes identical with that of a three-winding transformer. The values of reactance of the equivalent circuit can be obtained from the manufacturer of the machine.

In the final analysis the best way of dealing with excessive current is to raise the voltage. The generator can be connected directly to a transformer, both being treated as a single unit with differential protection around both, which is the equivalent of a high-voltage generator.

Unbalanced Faults

If a three-phase system is short-circuited across all three phases, it remains balanced, the calculation of fault current is in no way different from the calculation of load current, and we may use correctly the system voltage to neutral and the appropriate reactance values. If, however, the fault is of any other type—across two phases, two phases to ground, or one phase to ground—there results a distortion of phase voltages, which are no longer 120 degrees apart or equal in magnitude. The method of calculation of fault current used for three-phase faults is no longer valid.

For the selection of circuit breakers, the assumption of three-phase faults makes for easy calculations and is entirely satisfactory. In practice, however, three-phase faults are quite rare. The great majority are single-phase to ground and a few two phases to ground. If then we are faced with the problem of determining the true fault current, as in relaying or stability studies, the simple calculations of the foregoing examples are no longer sufficient.

The calculation of unbalanced faults, voltages, and currents uses the mathematical concept of "symmetrical components," by means of which any unbalanced system of phases can be converted into three balanced systems which, combined in appropriate manner, will give results rigorously equivalent to the unbalanced system. These three balanced systems are the positive phase sequence components, $A_1 B_1 C_1$, which are equal, are 120 degrees apart, and have the phase sequence ABC; the negative phase sequence components $A_2 B_2 C_2$, which are equal, are 120 degrees apart, and have the phase sequence ACB; and the zero phase sequence components $A_0 B_0 C_0$, which are equal and in phase.

Space does not permit the development of symmetrical components in this book, but some knowledge of their use is essential, and in the following chapters are taken for granted.

Application of Relays

The first point to consider in the application of relays is the current available on short circuit for operating them. The short-circuit current is closely tied in with operating conditions, especially in mixed water-power and steam-station systems, where the hydro stations, depending on water conditions, may operate as base-load stations; peak-load stations; or to provide wattless energy, or be shut down. The calculation of short-circuit current includes the determination of current flowing in all the branches under all these various conditions of generation.

Relays act to open appropriate circuit breakers on either side of the faulty machine or line section by energizing the breaker trip coils. A typical arrangement of protective zones is shown in Figure 8-9. Relays can be divided into three broad classes: those that protect against faults in individual machines and transformers, those that protect against faults in buses, and those that protect against faults in distribution and transmission circuits.

Machine Faults

The percentage differential relay described in Chapter 7 is the preferred relay for the protection of the system against faults in generators and transformers. Complete protection is obtained only when the neutral point of the generator is grounded, as otherwise there is no return path for the current in the case of a single-phase fault to ground. Where the generator neutral is grounded through a high resistance, as is done sometimes in unit-type stations, a sensitive ground relay in addition to the differential relay may be necessary. It is advisable to arrange the differential relay to trip not only the main circuit breaker and field switch but also the neutral breaker. Otherwise current will flow while the flux in the main field is decaying, and this flow may be enough to burn a hole in the iron.

The loss of generator field, which can lead to serious consequences if not protected against, is indicated by a flow of wattless current into the machine from the system when it loses its field. A watt-type relay

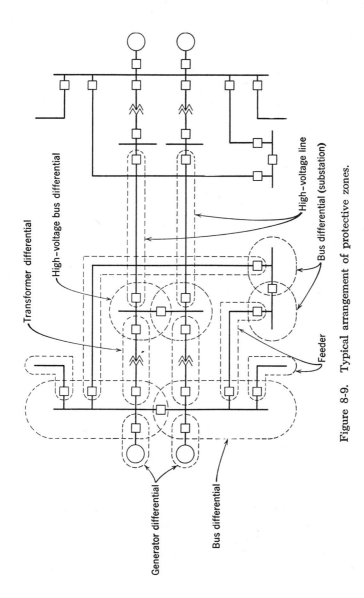

Figure 8-9. Typical arrangement of protective zones.

connected to measure reactive volt-amperes can be used for the purpose. The watt type or power relay is similar to the overcurrent relay except that the operating torque is obtained through energizing the upper coils from current transformers and the lower coils from potential transformers.

The reason that the generator draws wattless kva from the system when it loses its field is that it will slip poles and operate as an induction generator. If the other generators on the system have quick-response voltage regulators and the machine in trouble is not too big, they may be able to maintain system voltage; otherwise the voltage will sag and the whole system will fall apart.

Transformer Faults

The application of differential protection to transformers is not so simple as to generators for the reason that the current transformers on the high-voltage side and on the low-voltage side will have different ratios and may be of different types. Moreover, the inrush of magnetizing current which occurs when the transformer is energized will appear to the relay as an internal fault. This inrush current is due to the fact that it takes time to establish steady-state conditions in the transformer. The magnitude of the current, depending on the voltage at the instant it is energized, may be eight or ten times full load current at the first cycle, and the duration of the transient may be 5 to 7 cycles. The use of percentage differential relays is therefore even more necessary with transformers than with generators, and a less sensitive setting usually becomes necessary.

Most transformers are connected in star on the high-voltage side and in delta on the low-voltage side. This makes it necessary for the current transformers on the delta side to be connected in star, and on the star side in delta, so that the secondary currents of the current transformer will be in phase in the relay.

Bus Faults

It seems fairly obvious that differential relays can be used for protection against bus faults by balancing all the incoming currents against all the outgoing currents. The vector sum of all the secondary currents of the current transformers in the circuits to and from the bus is passed through the operating coil of an overcurrent relay in each phase. It will be zero under normal conditions. If there is a fault, however, the incoming and outgoing currents will not be equal, and a current proportional to the fault current will flow in the operating winding of the relay to trip all circuits connected to the bus. Differential protection applied to a bus is more complicated than to either a transformer or a generator. A number of current transformers are involved, and they are likely to have different ratio errors. Under such conditions the relay may operate even though the fault is external to the bus, particularly if the fault current is initially asymmetrical, because of saturation resulting from the d-c component. It becomes necessary then to determine the error current that may appear in the relay from the time the transformer saturates until the transient decays for the worst fault location of external fault; this is a tedious and complicated computation.

Transmission-Line Faults

The protection of systems against faults in transmission lines is obtained for the most part by the use of the distance relays described in Chapter 7. These relays are usually built with three protective zones, as indicated in Figure 8-10. The first zone covers 80 to 90%

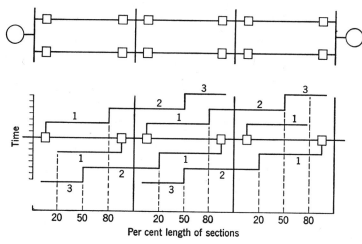

Figure 8-10. Protection by three-element distance relays. Zones of protection numbered 1, 2, 3.

of the protected section and clears faults in this section very quickly. The second zone is covered by a second distance element which has some time delay. It provides back-up protection for faults in approximately 50% of the adjacent section. The third zone covers the remainder of the adjacent line section to provide back-up protection for the first and second zones of the relay in the next section.

The distance relay has the disadvantage that it cannot be designed to cover the whole section to be protected. The reason for this is that the relay cannot distinguish between a fault just inside and just outside the protected section. There is therefore some 10 to 20% of the length of the section at either end which does not have the benefit of high-speed protection. This difficulty has been overcome by superimposing on the distance relay a carrier current signaling system by means of coupling capacitors. The carrier signal is used to block the tripping of the fundamental frequency relays in the unfaulted sections. In this respect it acts like a pilot-wire relaying scheme. Directional relays at the two ends of each section determine the location of the fault and control the carrier-current receiver unit to permit tripping if the fault is within the section, and to block tripping if the fault is beyond the section.

Since the carrier-current equipment functions only during a fault, the line can be used as a channel for telemetering and radio conversation. Moreover, if the carrier-current equipment fails it does not prevent the operation of the fundamental frequency relays in case of a line fault, but the tripping time will be increased.

In recent years microwave relaying has been tried out. The advantages claimed over carrier are freedom from interference and the fact that, because of independence of power lines, microwave channels can be established between points not suited to carrier, for example, from a dispatching office in a city to some outlying substation to which there is no overhead line.

Most transmission systems are operated with the neutral grounded. Since the majority of faults are to ground, ground relays are extensively used in addition to phase relays for the protection of transmission lines. Residual or zero-sequence current and voltage are used either alone or in combination to obtain reference quantities for the operation of ground relays. While distance relaying is by far the best means of relaying against phase faults, it is not always recommended for ground protection because of the wide variations in ground circuit impedance. Where overhead ground wires on steel towers are used and the impedance of the ground circuit is reasonably constant, impedance relays have been used successfully. In applications of this type, three relays using line-to-neutral voltage and line current must be used.

Grounding

There are two types of grounding:

1. Grounding equipment for safety purposes.
2. System grounding for the purpose of establishing a definite volt-

Power-System Fault Control

age zero point at the neutral point of a three-phase system, or at the middle point of a single-phase system.

The first type presents the problem of making sure that those parts of the equipment that are supposed to be dead are effectively grounded, so that, if there should be a leak to them from a conductor, there cannot be set up a voltage to ground which might endanger life or create a fire hazard. This type of grounding is governed by rules set down by the National Electric Safety Code of the Bureau of Standards and by the National Electrical Code of the National Board of Fire Underwriters. The NEC has been adopted by the American Standards Association as a national standard and has legal status in most cities and municipalities.

System grounding, on the other hand, is a problem of grounding electric circuits to safeguard service by making possible quick and reliable relaying. It also results in economy by reduction of the size of circuit breakers and system insulation and by improvement in system stability. It is the type of grounding with which utility engineers are mostly concerned.

The best ground, that is, the one with lowest electrical impedance, is any extensive underground water-pipe system, but frequently power stations and substations are located where there is no such water-pipe system. Even where there is one the authorities will not always permit connections to it.

The earth for the most part is a poor conductor of electricity—granite, for instance, is a good insulator. Because of the high resistivity, current flowing in the earth creates a considerable voltage drop, so that the potential of the earth is not necessarily zero.

Artificial grounds usually take the form of ground rods and ground mats. When a rod is driven into the ground the resistance of the earth immediately surrounding the rod is a maximum. With increasing distance from the rod the earth presents less and less resistance to the flow of current. Placing two or more rods close together is therefore not an effective method of reducing ground resistance. They should be spaced not less than 4 ft from each other.

The resistivity of the soil itself depends on its composition, degree of moisture, and contents of salts. It varies very quickly with these factors. For instance increasing the water content of a sample from 5 to 15% will reduce the resistance from 1000 ohms to 50 ohms, and adding 10% salt to the sample will further reduce the resistance to 5 ohms. It is important to drive rods below the frost line and if possible into permanently wet ground. The lower the station voltage, the lower should be the resistance of the ground mat. In a station with 100,000 kva fault energy and a one-ohm ground mat, the voltage drop through the mat will be 4200 volts at 13,800 volts and 420 volts at 138,000 volts.

Two factors enter importantly into the cost of the rods—diameter and length. The resistance varies very slowly with diameter, so that the chief consideration in selecting diameter is mechanical strength. The depth—that is, the length of the rods—reduces the resistance along an exponential curve, so that beyond 10 ft depth very little gain is obtained. Economically therefore the best rods to use in the great majority of cases are 0.5 to 1.5 in. in diameter and 6 to 10 ft in length. For best results they should reach below normal groundwater level.

In stations and substations ground mats are used rather than individual rods. Such a mat is made by driving the rods (usually of copperweld) into the ground in a rectangular pattern, the tops of the rods being connected together by copper cable (250,000 to 500,000 CM) some 2 ft or so below the surface. Such a mat may be 50 to 100 ft wide and 100 to 200 ft long. Rods are usually employed because they are cheap and easy to drive into the ground, but buried plates or straps make equally good grounds.

System Grounding

The general practice in the United States today is to ground the neutral of three-phase systems. In the lower-voltage distribution systems, 15 to 34.5, the neutral is not usually grounded directly to earth, but rather through some impedance such as a resistor, reactor, grounding transformer, or ground-fault neutralizer. In transmission voltages, 69 kv and up, the neutral is usually solidly grounded, although in the 69-kv class, neutral-grounding reactors of moderate ohmic size are sometimes necessary to limit fault currents.

In a perfectly transposed transmission circuit, the capacitances-to-ground of the three conductors are equal and displaced 120 degrees from one another. There will therefore be no potential between the neutral of the supply transformer and the neutral point of the three capacitances. Solidly grounding the neutral of the transformer, therefore, makes no difference to the voltage relations as long as the three-phase system remains balanced.

In a system covering a large area with a large number of transformers, it would be possible to ground the neutral of one transformer or of all transformers. Since no grounding resistors or reactors are used, either method could be considered to constitute "solid grounding." Electrically, however, it is obvious that the two extremes are not comparable since in the zero-sequence circuit the presence of one, two, or ten transformers in parallel makes quite a difference to the distribution and magnitude of fault current.

The effectiveness of the grounding of the system cannot, therefore, be expressed by a mere statement of what is done at the transformer neutrals, and the industry some years ago decided that the best method of defining the degree of grounding was to express it in terms of the ratios of zero sequence reactance and resistance of the system to its positive sequence reactance, X_0/X_1 and R_0/X_1. The definitions tentatively adopted read in part as follows.

Type A neutral-grounded systems are those which are well grounded, and the reactance and resistance ratios are less than for type B systems, but the system constants are not known in sufficient detail to establish the limiting ratios. (These are usually distribution systems.)

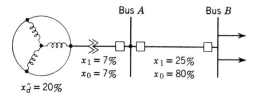

Figure 8-11. Example of determination of ground fault current in simple transmission system.

Type B neutral-grounded systems are those in which the reactance ratio X_0/X_1 is positive and less than 3 and at the same time the resistance ratio R_0/X_1 is positive and less than 1 at any place on the system. These limits correspond to the accepted definition of an "effectively grounded" system.

Type C neutral-grounded systems are those which are grounded but do not meet the requirements of the type B systems because the reactance ratio of 3 or the resistance ratio of 1 is exceeded, or both are exceeded. Systems using ground-fault neutralizers are included in this class. At points on the system where the ratio X_0/X_1 is 3 or less, the ground-fault currents will be of the same order as those at solidly grounded locations.

As an illustration of these definitions, assume a simple system shown in Figure 8-11.

At the bus A, the positive and negative sequence reactances of the system are each $20 + 7 = 27\%$. The ratio X_0/X_1 is $7/27 = 0.259$, and the system is therefore effectively grounded at this location.

Principles of Electric Utility Engineering

At the substation B, the positive and negative sequence reactances of the system are each $20 + 7 + 25 = 52\%$. The zero sequence reactance is $7 + 80 = 87\%$. The ratio $X_0/X_1 = 87/52 = 1.67$, and the system is effectively grounded at this location also.

At the station bus A, the three-phase short-circuit current is $100/27 = 3.7$ times full-load current. The fault current for a single-phase fault to ground is $I_g = I_1 + I_2 + I_0 = \dfrac{300}{27 + 27 + 7}$ = 4.9 times full-load current. To limit this single-phase fault current to a value equal to the three-phase fault current a reactor can be connected between the transformer neutral and ground. To obtain the desired value the zero sequence reactance must be increased to be equal to the positive sequence reactance, that is, 27%. The reactance must therefore be $(27 - 7)/3 = 6.66\%$.

The effect of this at station B is as follows: Without the reactor, the line-to-ground fault current is $I_g = \dfrac{300}{52 + 52 + 87} = 1.57$ times full-load current. With the reactor, $I_g = \dfrac{300}{52 + 52 + 107} = 1.42$ times full-load current. Thus it is seen that a reactor in the neutral of the transformer at the generating station equalizes the three-phase and single-phase fault currents without materially changing the line-to-ground fault current at the substations.

The ground-fault neutralizer or Petersen coil is a special form of neutral reactor. Its value is selected so that the current flowing through it for a line-to-ground fault is equal to the line capacitance current to ground that would flow if the reactor were not present. In this case the lagging current of the reactor and the leading current of line capacitance are practically 180 degrees out of phase, and therefore the actual current in the fault is substantially zero and the fault arc is extinguished. The combination of neutralizer reactance and line capacitance constitutes a parallel resonant circuit. If, however, the capacitances to ground should be unbalanced because of unsymmetrical line configuration, a zero-sequence voltage may exist between the neutral and ground. In this case the neutralizer and line capacitance constitute a series resonant circuit, and high voltages may result. This possibility must be watched in the application of these devices.

Small low-voltage systems are sometimes operated ungrounded. The advantage lies in the fact that such systems can be operated for a long time with one phase conductor grounded. Also, frequently

Power-System Fault Control

the faults clear themselves without interruption. As the system grows and the connected circuit miles increase, however, ungrounded operation becomes difficult because of arcing grounds and hazards to the public.

Generator Grounding

In a three-phase system or generator the normal rms voltage to ground is 58% of the line voltage. The test voltage of a generator is 1000 plus twice line-to-line voltage for one minute. The factor of safety is therefore 3.5 to 1 in a new machine. If the neutral is fully displaced, the factor of safety is reduced to 2 to 1. All machines even when old should be in condition to withstand this voltage.

An ungrounded generator is in effect a "capacitance-grounded" generator. Under normal conditions the charging currents flowing from the machine windings and leads to ground are sufficient to hold the neutral point close to ground potential, but, when a ground occurs on a phase conductor, a current flows whose magnitude is dependent on the size of the machine. It ranges from ¼ amp in a 10,000-kva, 2400-volt, two-pole generator to 12 amp in a 60,000-kva, 120-rpm generator. These currents are high enough to be dangerous and burn insulation if allowed to persist.

The mechanical bracing of the generator windings is designed to withstand stresses resulting from a 10-second three-phase short circuit at its terminals. If a generator is effectively grounded, the current through a faulted phase can considerably exceed the balanced three-phase short-circuit current. Limiting the current is of itself not sufficient to save the machine in case of a fault. A relatively small current, if permitted to persist, will melt a hole in the iron at the fault location. This fact explains why with all machines a sensitive relay scheme is necessary to take the machine out of circuit as quickly as possible if there is a ground fault.

The zero-phase sequence reactance of generators is less than the subtransient reactance. In order to limit the single-phase short-circuit current, therefore, to a value not exceeding the three-phase value, it becomes necessary to insert in the neutral of the machine some form of impedance, particularly where several generators are operating in parallel on a single bus. The impedance may take the form of a resistor or a reactor. With a number of generators on a single power bus, a common neutral bus is frequently used with the neutral grounding device between the bus and ground. Each generator neutral is connected to the bus through a breaker, and the breakers are interlocked so that not more than one can be closed at the same time. Where there are only two, or perhaps three, generators it may be more economical to provide each machine with its own grounding device.

In a station supplying feeders at generator voltage, preferred practice is to use a neutral resistor. With a reactor in the neutral the selection of the value is more critical as transient voltages of consider-

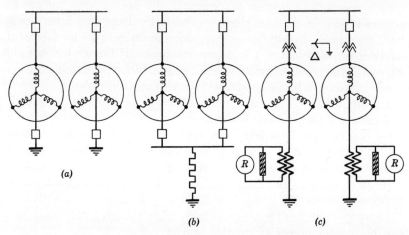

Figure 8-12. Methods of grounding generators. (a) Solid grounding. (b) Resistance or reactance grounding. In both (a) and (b) not more than one generator should be grounded simultaneously. (c) Grounding unit-type generation.

able magnitude can be set up during the fault. A neutral reactor adds to system reactance to give currents that lag the voltage by a large angle. At fault-current zero, therefore, the voltage may be near maximum and in a position to create oscillations involving the series reactance and shunt capacitance of the system. Here again the ratio of X_0 to X_1 should be not more than 3 if trouble is to be avoided.

Reactors cost considerably less than resistors of equal ohms, and for that reason they are used to some extent as indicated by the following figures taken from the 1947 *Report on Grounding Practice* of the AIEE. Of the systems using supply feeders at generator voltage, 15% were solidly grounded, 70% resistance grounded, 13% reactance grounded, 2% ungrounded, expressed as a percentage of the installed generating capacity.

In so-called unit-type stations the generator is connected directly to a transformer of the same capacity as itself, the combination making in effect a high-voltage generator. The transformer is delta-connected

Power-System Fault Control

at the generator voltage and wye-connected on the high-voltage side. This wye is usually solidly grounded at the neutral. The neutral of the generator may be grounded through a single-phase distribution transformer with a resistance in the secondary (Figure 8-12). During a line-to-ground fault the secondary resistor has sufficient loss to damp out transient voltages.

The 1947 *Report* mentioned above lists the systems transmitting power through transformers as being 78% solidly grounded, 6% resistance grounded, 12% reactance grounded, and 4% ungrounded on the high voltage side.

CHAPTER 9

Lightning Phenomena and
Insulation Coordination

It has been known for a long time that the earth is a negatively charged body and that the atmosphere surrounding it is a semiconductor of electricity. These facts led to the question of how the earth maintains its charge.

In 1920 C. T. R. Wilson, physicist at Cambridge University, offered an explanation which has since been confirmed by many observers and is now generally accepted. According to him, under fair weather conditions the negative leak from the earth is neutralized by a downward flow of positive ions so that a small voltage gradient—one volt per centimeter—is maintained near the surface of the earth. During a thunderstorm this balance is upset, and heavy negative currents flow in the earth. These heavy negative charges which compensate for the slow leakage to atmosphere are supplied by lightning strokes, of which a great number reach the earth every minute.

Each water droplet accumulates a negative charge as it falls through the electric field of the thundercloud, the gradient of which is much greater than that at the earth's surface—10,000 to 30,000 volts per cm. The upper part of the cloud is left with a positive charge which disperses and supplies the charge for the descending currents in the fair-weather areas.

Because the voltage gradient in the cloud is so much greater than that at the earth's surface, the lightning discharge is initiated at the cloud. The discharge starts as a pilot streamer which slowly ionizes a path through the air to earth. The pilot leader carries a charge with it; hence the potential gradient ahead of its tip is high, and breakdown of the air occurs. The leader proceeds toward the earth in jumps, at the same time charging the stroke channel. It is accompanied by secondary streamers branching out from the pilot

Lightning Phenomena and Insulation Coordination 161

leader. As the pilot nears the earth, the electric field becomes intense, and eventually a short upward streamer rises from the earth to meet the descending pilot. When they meet, an extremely bright return streamer propagates upward from the earth to the cloud, following the ionized path formed by the pilot leader. The charges distributed in the secondary streamers along the path are discharged progressively and give rise to the large currents associated with lightning strokes, which currents, as stated above, serve to restore the earth's negative charge.

When the initial pilot contacts the earth, we might expect the ionized path to act like a charged wire. It does not do so, however, because the path is not immediately able to pass any more current than it carried to lower the charge to earth, and the return streamer progresses up the ionized path at only one tenth the speed of light.

With the path established between the cloud and earth, secondary centers of charge in the cloud may spill over to the head of the ionized path and discharge directly to earth, giving rise to the phenomenon known as "multiple strokes." As many as twenty-two discharges have been recorded in such multiple strokes.

The small streamer that rises from the earth and initiates the stroke may start from any object connected to ground, such as a mast or tree or person. Four hundred people are killed and 1500 injured annually by lightning in the United States. The lightning damage is estimated at over 10 million dollars a year. To offset part of this loss, lightning has a commercial value in that it provides fertilizer. The stroke splits apart molecules of oxygen and nitrogen in the atmosphere which recombine into nitrous oxide. This in turn picks up free oxygen and water to form nitric acid, which the rain brings down to earth in a dilute solution that is beneficial. In this manner 1000 to 2000 lb of nitric acid are said to be deposited annually per square mile of the earth's area.

Surge-Measuring Instruments

In recent years lightning stroke currents have been measured in many localities. Statistical data on lightning stroke currents are necessary because the economical design of all equipment exposed to overhead lines is affected by the protection that must be provided against such currents. Laboratory oscillographs of various designs are the best instruments for determining the magnitude and wave shape of surge currents. Unfortunately, their cost and complexity make them unsuitable for field investigations.

The first useful field instrument was Peter's klydonograph, shown in Figure 9-1. It is cheap, simple, and generally applicable because it requires no skilled attention. It is based on the principle of so-called Lichtenberg figures. An electrode bearing upon a slowly rotating photographic film backed by a grounded plate produces a characteristic figure on the film whenever a voltage is applied to the electrode. Peters found from a large number of laboratory experi-

Figure 9-1. Klydonograph for measuring surge voltages. (Courtesy of Westinghouse Electric Corporation.)

ments that the figures could be correlated by their size and form not only with the magnitude of the surge but also with its nature: positive, negative, or oscillatory.

The next useful instrument was Faust and Kuehni's surge-crest ammeter, shown in Figure 9-2. This device consists of a small bundle of laminated permanent-magnet steel pieces. It is placed unmagnetized in the vicinity of the conductor, for example, a transmission tower leg, in which the current to be measured will flow. The remanent magnetism caused by the current is measured in terms of the current itself.

Both these instruments suffer from the same fault; they give no

Lightning Phenomena and Insulation Coordination 163

clue as to the shape and duration of the surge current. The need for this information led to the development of the fulchronograph (Figure 9-3), which is a refinement of the surge-crest ammeter. The laminations are mounted in the rim of an aluminum wheel, which is rotated at a constant known speed between coils through

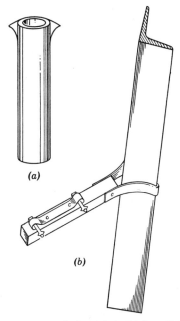

Figure 9-2. Surge-crest ammeter links. (*a*) Link. (*b*) Link in position on transmission-line tower. Links spaced 2 to 6 inches from the tower leg give a measuring range of 600 to 30,000 amperes.

which flows the surge current to be measured. The remanent magnetism, that is, the surge current, is determined as a function of time.

Several other instruments such as cameras, fusible wires, gaps, and surge integrators are employed for special purposes, but the ones described are the ones most commonly used.

Stroke Currents

From the information collected with these instruments, it has been determined that the great majority—perhaps 90%—of all lightning strokes are of negative polarity. In 75% of the strokes the current reaches its peak in less than 4 microseconds and declines to half value in 20 to 80 microseconds, with an average value of 43 microseconds. The current persists for periods ranging from a few

164　　　　Principles of Electric Utility Engineering

(a)

Figure 9-3. Fulchronograph. (a) Instrument mounted on pole for measurement of lightning arrester current. (b) Cover of fulchronograph removed. (c) Typical records of surge current on slow-speed and high-speed fulchronographs. (Courtesy of Westinghouse Electric Corporation.)

microseconds to fractions of a second; roughly half of all strokes exceed one tenth of a second, and 10% exceed one half of a second.

The magnitude of the stroke current to transmission lines has been recorded over a range of a few thousand to 218,000 amp.

Lightning Phenomena and Insulation Coordination

(b)

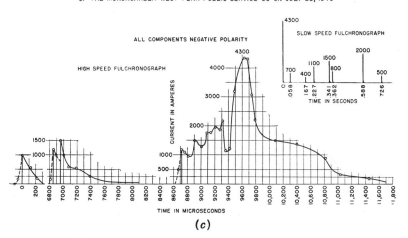

(c)

Figure 9-3 (*continued*).

Approximately 70% of all strokes exceed 9000 amp, 20% exceed 33,000 amp, and 5% exceed 63,000 amp.

Lightning discharges produce voltages in structures such as masts, trees, buildings, and transmission lines in two general ways, by direct stroke or by induction. If lightning strikes a structure, the lightning current flows in the structure and produces voltage.

Principles of Electric Utility Engineering

In a structure of high impedance, such as a tree, high voltages are developed. In conductors of low impedance, the voltage between electrically adjacent points is low. The high voltages in transmission lines result from the flow of lightning current into the distributed inductance and capacitance of the line. These are related by a quantity called the "surge impedance" of the line, Z in ohms. The exchange of energy for a free oscillation without attenuation is

$$\tfrac{1}{2}CE^2 = \tfrac{1}{2}LI^2$$

from which $$E/I = \sqrt{L/C} = Z$$

and the maximum possible voltage on suppression of E is IZ, where I is the surge current and Z the surge impedance.

The surge impedance for normally designed transmission lines is approximately 400 ohms. This results from the fact that the spacing between conductors increases with voltage so that the reactance and capacitance per mile of circuit remain more or less constant, the series reactance X_L being 0.8 ohm and the shunt capacitive reactance, X_C, 0.2×10^6 ohms. With these values

$$\frac{L}{C} = \frac{0.8}{\dfrac{1}{0.2 \times 10^6}} = 0.8 \times 0.2 \times 10^6 = 160{,}000 \quad \text{and} \quad \sqrt{\frac{L}{C}} = 400 \text{ ohms}$$

The corresponding value for cables is approximately 50 ohms. The voltage to ground in the case of a 10,000-amp stroke to an overhead transmission line would therefore be of the order of 4 million volts.

The stroke currents are in the nature of impulses. Hence the voltages are also impulses. There is no typical shape to these impulses. Usually there is only one peak, but there may be several. Lightning with a high current peak and short tail is called "cold lightning"; that with a low peak and long tail is called "hot lightning." Cold lightning causes explosive effects. It will seldom cause a fire. Hot lightning does not produce disruptive effects, but it will cause fires.

A lightning stroke will in general terminate at some object projecting from the surrounding earth surface. It has been observed that the frequency with which isolated masts or lightning rods on buildings are struck varies directly with height up to about 600 ft. Above that height the probability of a stroke increases faster, the reason being that the field at the tip of a very high structure becomes so intense that the upward streamer may actually initiate the stroke rather than the pilot streamer from the cloud.

Lightning Phenomena and Insulation Coordination

Lightning Protection

Several methods are used for protection against lightning. For such structures as oil tanks, buildings, munition dumps, etc., lightning rods, masts, and overhead wires are used to catch the lightning stroke and lead it to ground. The most complete protection is afforded when the protected object is surrounded by a grounded

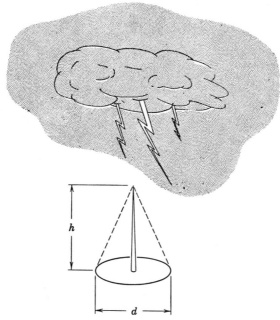

Figure 9-4. Protective cone provided by a grounded mast. With a mast of height h equal to the base diameter d, the exposure will be 0.1%, that is, one stroke in a thousand may be expected to reach the protected object.

shield. A mast will afford protection to a space around it which depends on the height of the mast, the space being in the form of a cone. Investigation has shown that protection within this space is practically complete if the base of the cone has a diameter equal to the height of the pole. (See Figure 9-4.)

In electrical supply systems protection is usually by means of lightning arresters and ground wires. About one half of one per cent of utility investment is in lightning protection equipment. Without this equipment outages due to lightning in the summer time would be intolerable. Accumulated data show that in any one year service interruptions to transmission lines in average lightning territory, if

168   Principles of Electric Utility Engineering

unprotected, may reach 2.3 per mile with a general average of 0.4. On a transmission line 50 miles long this rate would mean twenty outages per year.

Lightning arresters are in reality lightning diverters. They must provide an easy path to the ground for the lightning current without

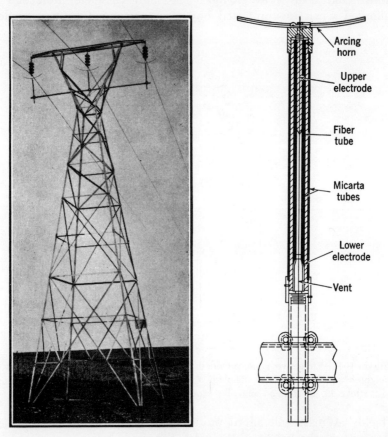

Figure 9-5. Expulsion-type arrester on transmission line. (Courtesy Westinghouse Electric Corporation.)

permitting voltages that endanger equipment and without permitting a service outage. An arrester is a very fast switch around insulation. It is normally open, but closes immediately when a transient voltage appears, and reopens quickly after the transient has disappeared, the closing mechanism being a spark gap. While in operation, the lightning arrester is also a path to ground for the system current, which it must also successfully cut off to prevent an outage. An ordinary

Lightning Phenomena and Insulation Coordination 169

gap (rod, horn, or sphere gap) would conduct lightning current to ground after flashing over, but such a gap would have no "cutoff" mechanism and "power-follow current" would flow until a line breaker operated. This would mean a system outage.

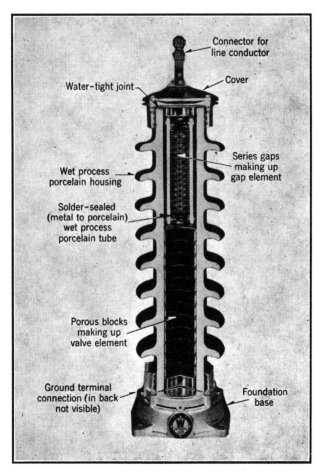

Figure 9-6. Valve-type lightning arrester. (Courtesy Westinghouse Electric Corporation.)

There are two broad classifications of arresters, the expulsion type, shown in Figure 9-5, and the valve type, shown in Figure 9-6. The expulsion type of arrester is essentially a tube, usually of fiber, with an electrode at either end. This provides a spark gap confined within the bore of the tube, the walls of which under the heat of the arc evolve a gas which blows the arc out through the bottom electrode.

The valve type has a series spark gap that normally provides insulation against the system voltage. When this gap flashes over because of a surge, power-fellow current is limited in value by means of a resistance material of special characteristics. The volts-amperes curve of a typical arrester is shown in Figure 9-7. The gap flashes over at a value of voltage which varies to some extent with the rate of rise of the surge current. For very steep waves the flashover will occur

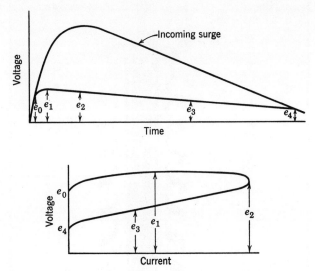

Figure 9-7. Typical performance curve of valve-type arrester.

$$\frac{e_1}{e_4} = \text{protective ratio}$$

at 4.5 times the arrester voltage rating, and for slowly rising surges at 3 times the arrester rating. When the gap has flashed over, the arrester commences to discharge surge current. When the surge current has passed through the arrester it cuts off the power-follow current at a voltage close to its rating. Thus the rating defines the maximum power-frequency voltage applied across the arrester terminals against which it can interrupt the power-fellow current and thereby restore itself to an insulator from the conducting condition into which it was put by the surge discharge. For instance, an arrester rated at 37 kv can be applied where the voltage across the arrester will not exceed 37 kv. This fact is important in hydroelectric systems, where the voltage may rise considerably if the load is dropped.

The breakdown of the gap in a valve-type arrester determines the

Lightning Phenomena and Insulation Coordination 171

initial discharge voltage, and the drop across the valve element determines the arrester voltage during discharge. These values and its ability to cut off the power-follow current are a measure of the arrester's protective ability. The ratio of the maximum surge voltage the arrester will permit to the maximum crest power voltage it will withstand following discharge is called the "protective ratio" of the

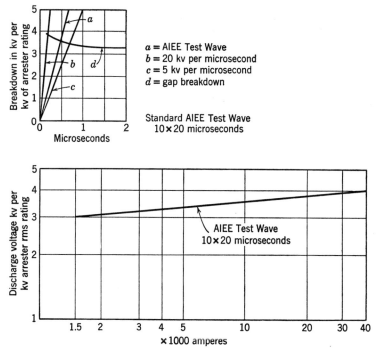

Figure 9-8. Typical discharge voltage characteristics for station-type arresters.

arrester. The fact that the protective ratio remains fairly constant through the range of voltage ratings of arresters means that the gap breakdown and the surge discharge voltage are approximately proportional to the voltage rating. The protective ratio will vary from 3 to 4.5, depending on the magnitude of the surge current and its rate of rise.

The sparkover voltage and discharge voltage for the standard AIEE test waves are shown in Figure 9-8. These values for a standard 73 kv arrester, for example, would be

$$\text{Sparkover: } 3.7 \times 73 = 270 \text{ kv in 0.5 microseconds}$$

$$\text{Discharge voltage: } 3.6 \times 73 = 263 \text{ kv}$$

Standard Wave Form

Natural lightning does not have any one typical base form. In general, the lightning-stroke current rises to a maximum in 1 to 4 microseconds, declines to half value in 20 to 80 microseconds, and has a total life up to several thousand microseconds. In order,

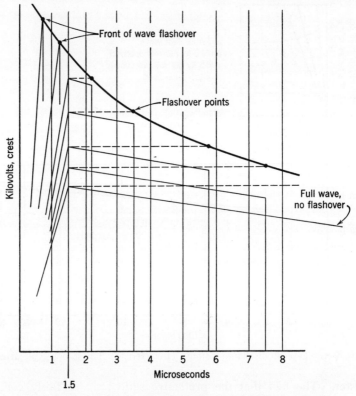

Figure 9-9. Typical volt-time curve of porcelain insulator.

therefore, to make tests and establish standards it was necessary to invent a standard wave, and the industry adopted as a standard test wave the so-called 1.5 × 40 wave. This is an impulse wave that rises to its maximum in 1.5 microseconds and falls to half its peak value in 40 microseconds. Impulse voltage tests are made with this wave. Impulse current waves through lightning arresters are made with a current wave of shape 10 × 20 microseconds. The reason for the difference is that the arrester elements influence the wave shape, and a (10 × 20)-microsecond current wave roughly approxi-

Lightning Phenomena and Insulation Coordination 173

mates a (1.5 × 40)-microsecond voltage wave at the arrester terminals.

The strength of insulation against impulse voltage is different from its 60-cycle strength. Generally speaking, the voltage required to flash over or break down insulation becames higher the shorter the time of application of the voltage or the steeper the rate of rise of the applied voltage. This is shown in Figure 9-9. If the crest voltages of the applied 1.5 × 40 wave are plotted against time-to-flashover, a curve is obtained which is called the "volt-time curve" of the piece under test. Different insulations have quite different time-lag characteristics, and on most insulations the volt-time curves are different for positive and negative impulses.

The high-voltage laboratories of the various manufacturers worked for several years establishing the flashover characteristics of all types of insulators in common use. As a result, there are available today standardized volt-time curves of gaps and insulators which can be used in the problem of insulation coordination.

Insulation Coordination

The extent to which arresters can limit the magnitude of transient voltages without damage to themselves determines to a large extent the insulation strength which must be built into the equipment on the system. The subject of correlating apparatus insulation and the protective devices to achieve this over-all protection is known as insulation coordination.

In 1930 a systematic investigation was undertaken by the utilities and manufacturers to establish insulation levels in the various voltage classes which would prevent promiscuous flashover of insulators and devices without putting an unnecessary economic load on the users. This led to ten years of intensive work in the laboratories to determine the flashover and puncture characteristics of all types of insulation used in the power industry. In 1940 the Committee submitted its report and recommended a table of Basic Impulse Insulation Levels, commonly referred to as BIL's for all voltage classes. The report was generally accepted and adopted by the industry.

The basic impulse insulation level is defined as "a reference level expressed in impulse crest voltage with a standard wave not longer than (1.5 × 40)-microsecond wave. Apparatus insulation as demonstrated by suitable tests shall be equal to or greater than the basic insulation level." The values of BIL decided upon for the various voltage classes (as amended) are given in Table 9-1. It will be noticed that for systems above 69 kv two values of BIL are given.

The reason is that at these higher voltages there is an economic gain in making a distinction between ungrounded and grounded systems. In the latter, since the voltage to ground is more or less stabilized, the insulation level requirements are lower. When reduced insulation is used, however, more care must be exercised in the

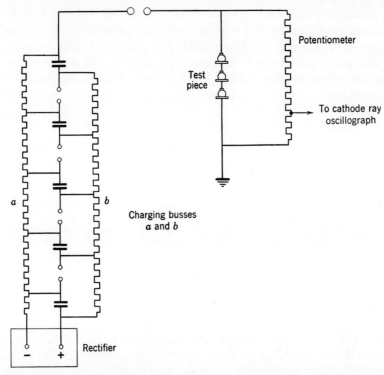

Figure 9-10. Principle of the Marx surge generator. Voltage is applied across busses *a* and *b* to charge the capacitors. At some predetermined value the gaps flash over and the capacitors discharge in series across the test piece.

application of the protective equipment. Studies are continuing to determine the minimum values of BIL for operating voltages above 230 kv which will result in the best over-all economy.

The impulse generator used to carry out the large amount of surge testing that became necessary to develop insulation coordination properly is the one devised by Dr. Emil Marx in Germany. It consists essentially of a group of capacitors, spark gaps, and resistors so connected that the capacitors are charged in parallel from a relatively low voltage source and discharged in series to give a high voltage across the test piece. The principle is shown in Figure 9-10.

Lightning Phenomena and Insulation Coordination

TABLE 9-1. STANDARD BASIC IMPULSE INSULATION LEVELS

Reference Class, kv	BIL, kv Crest	Reference Class, kv	BIL, kv Crest	
1.2	30	69	350	
2.5	45			
5.0	60	115	550	450
8.7	75	138	650	550
—	95	161	750	650
15	110			
23	150	230	1050	850
34.5	200			
46	250			

A potential divider supplies a reduced voltage to the cathode ray oscillograph proportional to the test voltage. The shape of the impulse wave applied to the test piece is determined by the constants of the discharge circuit, some of which are inherent in the capacitors and leads, and some are added externally.

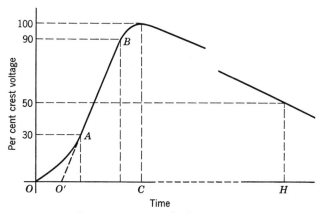

Figure 9-11. Standard test wave.
$O'C = 1.5$ microseconds; $O'H = 40$ microseconds.

The front of the wave is not a direct proportionality between voltage and time, but rather it takes the form shown in Figure 9-11. For practical purposes, therefore, an arbitrary front has been adopted as standard, which is represented by a straight line drawn through the 30 and 90% points of the rising voltage—points A and B in Figure 9-11. The standard (1.5×40)-microsecond wave has a front equal to 1.5 microseconds $O'C$ and a time to half-crest $O'H$ equal to 40 microseconds.

The three types of insulation that power system engineers are

176 Principles of Electric Utility Engineering

interested in are (1) air, as represented by spacing between conductors and buses, also by rod gaps; (2) porcelain, in the form of insulators of various shapes and sizes; and (3) composite insulation such as is used in machines, transformers, and cables.

Flashover in air varies with temperature, barometric pressure, and humidity. Standard correction factors for these have been developed. Rod gaps are used around bushings to establish definite flashover values, particularly where local conditions, such as salt fog and dirt, make oversize bushings desirable. In a transformer, for instance, it is obviously desirable that the bushing flash over at a lower value of voltage than that required to puncture the insulation.

Porcelain insulators have been improved over the years to the point where quite generally the flashover voltage is lower than the puncture voltage. Unless an arc persists so long that the insulator fails from thermal effects, the insulator will not be damaged by flashover.

Of the heavy equipment with composite insulation in power systems, the transformer is the one most frequently exposed to lightning and switching surges. Moreover, the cost of a transformer depends largely on the amount of insulation built into it. For that reason, the behavior of transformer insulation against surges has received intensive study. The insulation in question is the main insulation to ground and the minor insulation between turns. As might be expected, the puncture characteristics of such insulation differ from the flashover characteristics in air. Coordination between the winding insulation (which punctures) and the bushing (which flashes over), therefore, requires special care, as at very short times, up to 3 or 5 microseconds, the insulation might fail before the bushing flashed over.

The 138-kv system shown in Figure 9-12 serves to illustrate the application of insulation coordination. The substation has two incoming lines supplying two star-delta transformers through two circuit breakers. The transformers have their neutrals effectively grounded. The line insulation consists of nine suspension insulators and the bus insulation in the substation of twelve similar insulators.

The volt-time curves of all this equipment may be obtained from handbooks or from the manufacturers and plotted together with the BIL line on cross-section paper. The BIL for 138 kv is 650 kv. All the volt-time curves lie above the BIL line, since by agreement they must all be "equal to or better than the BIL." The maximum surge that nine suspension insulators will permit to enter the station without themselves flashing over is found from their volt-time curve to be 860-kv crest.

Lightning Phenomena and Insulation Coordination 177

A traveling wave coming into the station is limited in magnitude at the arrester location to the discharge voltage of the arrester, and a wave of this magnitude (and with the same rate of rise as the original wave) travels on to the station terminus and is reflected.

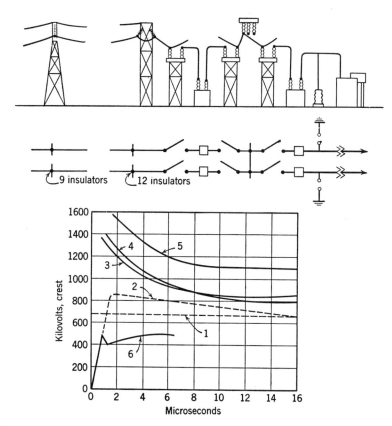

Figure 9-12. Application of insulation coordination. (1) Basic insulation level and transformer insulation. (2) Maximum 1.5 × 40 wave permitted by nine insulators. (3) Line insulation. (4) Disconnecting switches. (5) Bus insulation. (6) Arrester.

If the terminus is a dead end, the reflected wave will be twice the arrester voltage; if the terminus is a transformer, it will be somewhat less than twice. This possible reflection must be taken into account in the application of arresters. Since the system is effectively grounded, an arrester of the next lower voltage class (115 kv) may be used. Such an arrester is rated at 121 kv (115 + 5%).

For a traveling wave coming into a dead-end station the discharge current in the arrester will be

$$I = \frac{2E - V}{Z}$$

where E = the surge voltage,
V = the arrester terminal voltage,
Z = the surge impedance of the line.

The arrester gap breakdown for the standard AIEE test wave is 3.6 kv per kv of arrester rating (121 kv) or $3.6 \times 121 = 435$ kv $= V$. (See Figure 9-8.)

The current to be handled by the arrester with a standard wave will therefore be in the example cited:

$$I = \frac{(2 \times 860{,}000) - 435{,}000}{400} = 3250 \text{ amp}$$

From the curve in Figure 9-8 the discharge voltage of the arrester for 3250 amp is 3.2 kv per kv of arrester rating, or $3.2 \times 121 = 388$ kv, which value will be added to the chart as the protective level of the 121-kv arrester.

The application of an arrester in practice is not so simple as this example would imply. Protection against surges is not an exact science. The rate-of-rise and magnitude of the incoming surge will affect the protection as will many other factors, such as the number of circuits connected to the bus, the shielding provided, the distances from transformer to arrester and from arrester to true ground. With so many variables we must fall back largely on statistical probability in the area under consideration.

The location of the arrester in relation to the equipment to be protected is important. The electrical distance between them should be as short as possible. It is also important that the leads from the arrester to ground be as short and direct as possible, because with the heavy currents involved the voltage drop in these leads can be quite high and add directly to the arrester discharge voltage. Consider an arrester protecting a transformer as in Figure 9-13. V is the voltage across the arrester, e_ω is the inductive drop in the lead $\left(= L\frac{di}{dt} \right)$, e_t is the voltage to ground at the transformer terminal ($= V + e_\omega$). The inductance of a straight conductor is of the order of 0.4 microhenries per foot. If the surge current rises at a rate of 5000 amp per microsecond, the voltage drop in the lead will be

Lightning Phenomena and Insulation Coordination 179

2000 volts per foot, or in a 20-foot connection 40,000 volts, a value which cannot very well be ignored.

The ground resistance must also be taken into account. If the arrester and transformer have separate grounds, there is an additional IR drop which adds to the voltage e_t. The effect of this

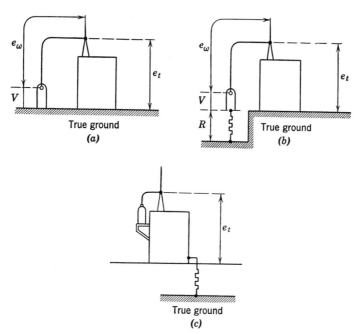

Figure 9-13. Arrester mounted on transformer. *Example:* 37-kv arrester, 30-ft lead, 5000-amp stroke, $R = 5$ ohms.

Discharge voltage $V = 37 \times 3.6 = 133$ kv
Lead drop $e_\omega = 30 \times 5000 \times 0.4 = 60$ kv
IR drop $= 5000 \times 5 = 25$ kv
(a) $e_t = 133 + 60 = 193$ kv
(b) $e_t + IR = 193 + 25 = 218$ kv
(c) $e_t = 193$ kv but 25 kv above true ground

ground resistance, however, can be eliminated if the arrester ground terminal is connected to the transformer tank. The potential of the tank and arrester may be raised above true earth, but the potential around the transformer insulation is limited to e_t. The practice of mounting the arrester on the transformer tank is frequently adopted where the transformers are designed with "reduced insulation," that is, for an insulation level lower than that normally associated with the system voltage. The cost of a high-voltage transformer is materially influenced by the impulse insulation level for which it is built. For instance, a transformer of, say, 50,000 kva, 138 kv, will cost 10% less if it is designed with a BIL of 550 kv instead of the 650 kv normally associated with 138-kv systems. However, this lower BIL can be used only if particular care is taken that a satisfactory arrester is installed and is properly located.

Machine Protection

Surges coming into a substation on an overhead line may produce over-voltages on the low-voltage side of a transformer by electrostatic and coupling. Under certain circumstances it is advisable, therefore, to protect rotating machines connected to overhead lines even though a transformer intervenes. The impulse strength of machine insulation because of space limitations is of necessity not much better (perhaps 25%) than the peak of its 60-cycle test voltage when new.

The stress on the insulation to ground is determined by the magnitude of the surge voltage. The stress on the turn insulation is a function of the rate of rise of surge voltage. Therefore, protection must take the form of not only limiting the surge voltage but also sloping the wave front. This is done by a combination of lightning arrester and capacitor installed at the machine terminals. The capacitor is selected to slope the incoming wave to not less than 10 microseconds to crest. The arrester used is a special design with low protective ratio. To make the capacitor effective requires that a second arrester of normal design be installed some distance ahead of the machine to be protected. If the machine is connected directly to the overhead line, as may be the case in mine installations, for instance, the distance from machine to the far arrester may be 500 to 1500 ft. If a transformer intervenes, the arrester on the high voltage side of the transformer will suffice.

CHAPTER 10

Transmission Systems

The principal elements that go into a transmission system having been reviewed, some of the economic aspects may now be considered. There is no clear-cut answer to the question: "How should X kilowatts be transmitted from source A to load-center B?" The variables are innumerable. The first question may well be: "Should the transmission be undertaken at all?" In certain circumstances it may be cheaper to haul coal over railways or to pump fuel oil through pipe lines than to transmit electric energy over transmission lines.

Studies of the comparative costs of transporting energy by different methods have been made frequently since it was suggested in 1890 that the energy of Niagara Falls be transmitted to Buffalo by compressed air. The latest of these studies is one to be found in the 1948 *AIEE Transactions* by R. E. Pierce and E. E. George, of Ebasco Services, Inc. In order to make a comparison between transportation of coal, gas, oil, and electric transmission, the authors plotted the cost of delivering one kilowatt-hour against distance. The result is shown in Figure 10-1. They used for this purpose the following conversion factors:

$$1 \text{ barrel of oil} = 6{,}250{,}000 \text{ Btu} = 500 \text{ kwhr}$$
$$1000 \text{ cubic feet of gas} = 1{,}000{,}000 \text{ Btu} = 80 \text{ kwhr}$$
$$1 \text{ ton of coal} = 25{,}000{,}000 \text{ Btu} = 2000 \text{ kwhr}$$
$$1 \text{ kilowatt-hour} = 12{,}500 \text{ Btu, i.e., } 27\% \text{ efficiency}$$

The authors point out that the comparisons are of necessity only approximate because of the differences in costs of the fuel at their sources.

To obtain the cost of electrical transmission, including fixed charges, losses, operation, and maintenance, the authors used for 90% load factor the formula

$$\text{mills/kwhr} = 0.30 + \frac{(0.35 \times \text{miles})}{100}$$

182   Principles of Electric Utility Engineering

Where hydroelectric power is involved, a transmission line is nearly always a necessity. With a thermal plant, the requirement of a plentiful supply of fuel for years to come and the immediate requirement of a large volume of cooling water for the condensers must be balanced against distance from the load centers. The problem of the best location involves real-estate values, fuel transportation costs,

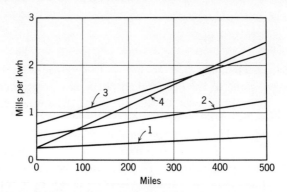

Figure 10-1. Cost of transport of energy. (1) Oil by pipe line. (2) Gas by pipe line. (3) Coal by railroad. (4) Electricity—90% load factor.

river flow and temperature variations, right-of-way costs, and many other factors. But if an overhead transmission line has been decided upon, the first choice to be made as far as the line is concerned is that of voltage. The voltage selected will be the one already in general use in the area for the type of transmission being considered. For instance, if the utility, or a neighboring utility, already has a considerable network of 230 kv lines, the management is not likely to adopt 161 kv for a new circuit even though this voltage may prove cheaper initially.

The importance of the new circuit will be a deciding factor in determining how much money can be invested in it. Can the system stand an outage of this line? If so, for how long? If the transmission line consists of two circuits, how close to the stability limit can we go with one circuit or section of circuit relayed out in case of a fault? What additional investment is justified to make the line lightning-proof? The right-of-way, depending on whether it goes through open country or through a semi-urban district, also can become a deciding factor in the design of the line. All these questions have a marked influence on the cost, but they cannot be included in any single formula. For this reason Kelvin's law, which is applicable to short lines where current-carrying capacity and losses are controlling, is not readily usable in the economic study of long-distance transmission at high voltages.

In the past, steel towers proved more economical than wood structures for the majority of high-voltage lines, 115 to 230 kv. But with changes in economic values and availability of materials there has been a trend to wood so that most of the lines built in recent years in the 115- and 138-kv classes have used wood poles. The longer life of wood poles due to improved and cheaper methods of treatment against fungi and insects is a factor that has contributed to this result, as has the higher cost of steel towers. In the 230-kv class because of the heavier loadings steel towers are still the preferred material, but even here wood structures are used in a few cases where local conditions make them preferable.

Span and Sag

As has been indicated, the choice of the type of supporting structure—wood or steel—depend on local conditions. This decision having been made, the next factor to consider is the span, that is, the distance between supporting structures. The choice of span is not always entirely in the hands of the engineer. Topography will frequently leave little choice in the location of the towers. If we assume a fairly flat terrain, however, the choice of span becomes a matter of economics.

The stress in the conductor is the parameter on which everything else is based, and the sag in the conductor as it hangs between adjacent towers determines the stress. Since the stress depends on sag, any span can be used provided the towers are high enough and strong enough. The matter is merely one of extending the catenary in both directions. But the cost of the towers goes up rather steeply with height and loading, and the problem becomes the balancing of a larger number of lighter towers against a smaller number of heavier towers. Fortunately, if the total cost per mile is plotted against span length, there results a well-defined minimum point to give the most economical span. This normal span usually runs from 500 to 900 ft where wood structures are used, and from 700 to 1200 ft where steel is used. However, no transmission line has all its spans of the same length, and spans up to three times normal are to be expected. The sags vary from 2 to 6 ft per 100 ft of span.

The loading on the supporting structures consists of the stress in the conductors acting longitudinally, the weight acting vertically, and the wind acting transversely. It is common practice to limit the tension in the conductor (without wind and ice) to 25% of its

184   Principles of Electric Utility Engineering

ultimate strength. It has been found that when wires are heavily stressed they fail in a relatively short time because of fatigue brought on by vibration. This vibration, which is of the order of 15 to 100 cycles per second, is set up by light transverse winds creating eddies on the leeward side of the conductor. The length of span and the working stress appear to have a marked influence on this type of failure, and some experts feel that 25% loading is too high. The vertical load on the tower is due to the weight of the conductor, to which must be added the weight of a coating of ice. The transverse load depends on the assumed velocity of the wind.

The maximum wind velocity, temperature, and ice formation of course vary in different parts of the country. The National Electric Safety Code of the National Bureau of Standards divides the country in three parts: heavy-loading, medium-loading, and light-loading areas. The heavy-loading area comprises roughly the northeast quarter of the

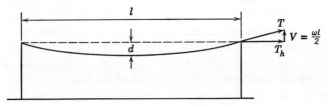

Figure 10-2. Approximate method of determining sag.

country; the medium-loading area comprises the northwest quarter plus a strip across the middle of the country; the light-loading area comprises California and all the southern part of the country to a depth of 300 or 400 miles. The values suggested in the Code are useful as a guide even though they may have to be modified to meet local conditions.

The maximum stress in the tower generally results from the specified loads acting simultaneously, but in some designs maximum load comes from the torsional stresses resulting from one or more broken conductors. Calculations are usually based on failure of one conductor. Various designs of flexible towers have been suggested from time to time to deal with the problem of excessive stress in the towers due to broken conductors.

The conductor between the towers forms a catenary, but for the small ratio of sag to span ordinarily used there is very little difference between a catenary and a parabola. In determining the sag we may therefore use the formula for a parabola, which is simpler than that for a catenary. In Figure 10-2, T is the tension in the conductor at

Transmission Systems

the tower, T_h the horizontal pull, V the vertical loading resulting from the weight of the conductor and ice, all in pounds. l is the span and d the sag in feet. If the weight of the conductor is ω pounds per foot, then

$$d = \frac{l^2 \omega}{8 T_h}.$$

One may substitute $\alpha \times$ area for ω, where α is pounds per foot per square inch, and $f \times$ area for T, where f is the stress in pounds per square inch. The formula is thus simplified to

$$d = \frac{l^2 \alpha}{8f}$$

As an illustration, assume a copper cable conductor of 300,000 CM (0.236 sq in.), weight 4890 lb per mile, breaking strength 13,170 lb, suspended between towers 1000 ft apart, and determine the sag.

$$\alpha = \frac{4890}{5280 \times 0.236} = 3.95 \text{ lb/ft/sq in.}$$

With 25% loading,

$$f = \frac{13,170}{4 \times 0.236} = 13,950 \text{ lb/sq in.}$$

$$d = \frac{1000 \times 1000 \times 3.95}{8 \times 13,950} = 35.5 \text{ ft}$$

This sag is only approximately correct for the reason that the formula for a parabola was used and the total stress T instead of the horizontal component T_h, but the error is probably well within 5%. Where the towers are at different elevations, the same equation of the parabola may be used unless the difference in elevation is considerable, as in coming down a mountainside. In such cases the assumption of a parabola is not valid.

Erecting or "stringing" the conductor on the towers to obtain the correct sag and conductor tension is complicated by the fact that the spans are of necessity not all always the same. To simplify the work, so-called sagging charts or tables are used. Of these the Thomas charts and the Martin tables are the best known.

Power Transmitted

The power that can be transmitted, that is, with a given voltage the amperes in the line, determines the cross-section of the conductor, but its diameter is dictated by the corona limitations previously discussed (Chapter 6). Because these two requirements do not always coincide, we have such peculiar conductors as the HH conductor made up of interlocking segments stranded together to form a flexible tube; the I-beam conductor in which the wires are laid helically over a twisted I-beam; hemp-filled aluminum conductors, and the like. A stranded conductor of the required diameter made up entirely of wires not only has a much greater cross-section than necessary to carry the current, but also its greater weight adds unnecessarily to the over-all cost. For example, a one-inch diameter stranded copper conductor weighs 13,000 pounds per mile and carries 1100 amp. A one-inch tubular conductor may weigh only 4050 lb per mile and carry 700 amp, or 8060 lb per mile and carry 900 amp.

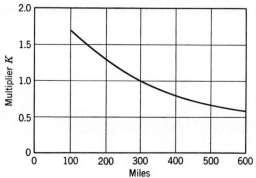

Figure 10-3. Transmission capability of conventional lines. kw = $K(2.5\,\text{kv}^2)$.

The power that can be transmitted may be conveniently based on the surge impedance of the line. In a paper presented before the AIEE in January, 1951, H. P. St. Clair and E. L. Peterson of the American Gas and Electric Corporation published a curve of megawatt capability of lines based on this concept. The surge impedance of a line is the value $\sqrt{L/C}$, where L is the reactance of the line and C the capacitance. It is of the order of 400 ohms for normally designed transmission circuits. The surge impedance loading is the unity power-factor load at which the I^2X of the line equals its charging kva. It is equivalent, therefore, to a load impedance of 400 ohms from line to neutral. Expressed in this manner, the loading is independent of the voltage, and this fact enabled the authors to draw a curve of transmission capability against length of line, which for all conventional lines using single conductors per phase is

$$\text{kw} = \frac{1000\,\text{kv}^2}{400} = 2.5\,\text{kv}^2$$

Transmission Systems

Thus the curve, which is reproduced in Figure 10-3, applies to any voltage. The authors state that for lines 300 miles and above the loads shown are based upon reasonable assumptions of generator and transformer impedances, with a reasonable margin below the steady-state stability limit to allow for transient load swings. For lines below 300 miles further allowance was made to include real and reactive losses. Table 10-1 gives megawatt capability of lines for various voltages and distances. The charging kva of a line per 100 miles of circuit is approximately equal to 20% of the surge impedance loading at 300 miles.

TABLE 10-1. MEGAWATT CAPABILITY OF THREE-PHASE CIRCUITS

Line to Voltage, kv	Distance, miles			
	300	200	150	100
115	33	43	50	55
138	48	63	73	82
161	65	85	97	109
230	132	172	198	222

Conductor Configuration

When the type and size of conductor have been determined, their configuration must next be considered. A single-circuit high-voltage line is likely to have a flat configuration of conductors. A twin-circuit line, or a single-circuit line to be duplicated in the near future, is more likely to have a vertical arrangement. Formerly the spacing between conductors was based on the operating voltage and sag. In the flat configuration it was assumed that the conductors in the middle of the span in a high wind might swing in opposite directions if their natural period of oscillation differed because of slight differences in sagging. In these circumstances any two conductors should remain far enough apart to prevent flashover between them. In the vertical arrangement the conductors were hung so that, if one spilled its load of ice, it would not spring up at midspan into the conductor above it. For this reason the middle cross-arm was (and still is) made a foot or two longer than the top and bottom cross-arms.

These are still important considerations, but with better knowledge of lightning protection more attention is given to prevention of flashover of a conductor to ground during lightning storms. The basic principles underlying the design are:

1. Ground wires must be located to shield the line conductors from direct strokes.

2. Adequate clearance from conductors to towers or to ground

must be maintained at all times (even during maximum swing due to wind, or maximum sag due to sleet) so that the flashover voltage through air is at least of the insulator string.

3. Adequate clearance must be maintained at midspan to prevent flashover from ground wires to conductors.

4. Tower-footing resistance must be as low as economically justified.

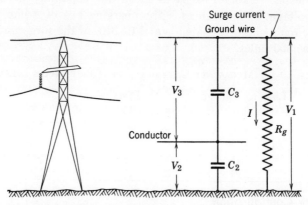

Figure 10-4. Relationship of voltages on transmission line during lightning stroke.

V_1 = voltage drop through tower
V_2 = voltage of conductor to ground
K = coupling factor conductor to ground wire
V_3 = voltage across insulator string = $V_1 - V_2 = V_1(1 - K)$

When lightning strikes a transmission line, the stroke is expected to impinge on the ground wire and the lightning current flows along the wire, down the tower, and through the tower-footing resistance to ground. The entire tower top and ground wire rise to a high potential because of the IZ drop through the tower and ground resistance, the line conductors being held close to ground potential by virtue of their capacitance to ground. If there were no coupling between ground wires and conductors, the full potential at the tower top would appear across the string of insulators. However, the coupling factor reduces the voltage stress because the induced potential on the conductors is of the same polarity in relation to ground as the lightning stroke. The coupling factor is of the order of 30 to 40%. The relationship of these voltages to one another is shown in Figure 10-4.

The shielding of a conductor by a ground wire depends on its location above the conductor. It need not be, and usually is not, directly

Transmission Systems   189

above the conductor. The action is similar to the "cone of protection" discussed in Chapter 9, except that in this case it becomes a "wedge of protection," with the ground wire forming the sharp edge of the wedge. The angle of protection, that is, the angle formed by a line through the ground wire and conductor, and the vertical should not exceed 30 degrees. Where two ground wires are used, as is usual with flat configuration, the middle conductor has little exposure, and the angle of protection can be much greater than 30 degrees.

Flashover of a transmission circuit is admittedly a random phenomenon. A thunderstorm in the vicinity of a line is no indication that the line will be struck, but the probability of its being struck goes up with the frequency of the storms. The number of storms annually in any given area is known as the "isokeraunic level." Isokeraunic maps for the United States are published by the Weather Bureau. The isokeraunic level for the greater part of the country east of the Rocky Mountains lies between 30 and 50. In the states along the West Coast and along the Canadian border the level drops to 10 and 15. This explains why Pacific Coast utilities for the most part do not use ground wires on their transmission lines. There are two locations where the frequency of storms is much above the average, Florida with 80 to 90 storms and New Mexico with 60 to 70 storms. Field studies have established the probability of lightning strokes attaining certain values of severity so that, with the tower-footing resistance and the coupling factor given, it is possible to determine the surge current required to flashover a given string of insulators. The probability of flashover is the same as the probability of tower currents equal to or greater than this value. E. L. Harder and J. M. Clayton in a paper presented at the winter meeting of the AIEE in 1949 published a series of curves based on this principle. They show how many standard insulator units (10 × 5¾ in.) should be included in a string of insulators supporting a conductor to keep the probability of flashover down to any desired level. The number of insulators is plotted against outages per 100 miles of line per year, and corresponding stroke current in amperes for various tower-footing resistance values from 0 to 100 ohms. The family of curves include spans from 200 to 2000 ft. Figure 10-5 shows, as an illustration, the curve for 1000-ft span and 10 ohms footing resistance. From this curve it is seen that, if the performance is to be two outages per 100 miles per year, the insulator strings have to be made up of seven units, and the line should be good for strokes up to 85,000 amp. The midspan clearance must of course be consistent with the insulation at the tower. Curves correlating the two are also given.

Experience has shown that the predictions of line performance against lightning are fully realized. In other words, it is possible to design a transmission line today with a fairly certain expectation that the outages due to lightning will be as predicted. The transmission lines built from Hoover Dam to the vicinity of Los Angeles by the Metropolitan Water District, the Southern California Edison Company, and the City of Los Angeles offer a unique opportunity to

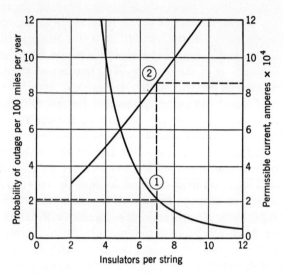

Figure 10-5. Probability of outage ① and permissible stroke current ② for 1000-ft span and 10-ohm tower footing resistance.

compare performance because, although quite different in design, they traverse the same general terrain with an isokeraunic level of about 15. The results were reported in *Electrical World*, June 18, 1949.

The Metropolitan Water District of Southern California operates a 230-kv line consisting of one circuit of 795,000 CM ACSR conductors in flat configuration on conventional towers with provision for two ½-in. steel overhead ground wires. The insulator strings consist of thirteen porcelain suspension units 10 × 5¾ in. Initially, overhead ground wires were omitted except for sections adjacent to the stations. The line was placed in operation in November, 1938. During the first four years with no overhead ground wires, the service interruptions from lightning averaged twenty-seven per year. The reason these outages could be tolerated is that the load was entirely a pumping load, and interruptions of short duration had no influence on the

Transmission Systems 191

volume of water pumped. With the addition of load that required uninterrupted service it became necessary to add the ground wires. This was done in 1943. Since that time there have been three outages in 6 years.

The line of the Southern California Edison Company consists of two circuits on separate steel towers with flat configuration. Each circuit is supplied at Hoover from a step-up transformer bank of 150,000 kva at 220 kv. The line is 233 miles long to the outskirts of Los Angeles. The ground wires are midway between conductors, 8 ft above them at dead-end towers and 12 ft above them at suspension towers. The midspan clearance is 25 ft. The ground wires are ½-in. high-strength steel. A number of towers were tested for tower-footing resistance, and fifty-four tower locations were equipped with two radial counterpoises. Each of these consisted of four 200-ft lengths of ½-in. galvanized steel cable buried and tamped in a trench 1.5 to 2.5 ft deep. The tower-footing resistance was lowered in this manner to an average of 9.3 ohms. The lines have had twelve outages since 1939 due to lightning or about 0.25 outage per 100 miles per year.

The third transmission line, put up by the Los Angeles Department of Water and Power, consists of three circuits operated at 287 kv. The lines were designed for high reliability and carry two ground wires and two continuous counterpoise wires. Tower-footing resistance measured 1 to 150 ohms before installation of the counterpoises and 0.4 to 10 ohms afterwards, with an average of 0.8 ohm. Stroke indicators show that the circuits have been struck by lightning at least 200 times. In 12 years of operation, only three flashovers have been attributed to lightning, but even these are questionable.

Counterpoises

The counterpoises mentioned above are wires buried 2 ft or so underground and connected to the towers to reduce their impedance to ground. They come in two forms, the so-called crow's foot, in which the wires extend radially from the tower, and the parallel type in which the wires run parallel to and under the line conductors. As the name implies, a counterpoise is supposed to reduce the coupling between the conductors and ground wire by pulling down the potential of the ground wire toward true ground. If coupling were the predominating factor, then for a given amount of wire it would be advantageous to use the parallel type of counterpoise. If ohmic leakage resistance predominates, then the crow's foot counterpoise should be more effective. Tests have shown that the coupling between counterpoise and ground wire is small—3 to 8%—and that counterpoises have been successful primarily by reducing leakage resistance to ground. This would indicate a preference for the crow's foot type, but, despite this, parallel counterpoises are more commonly used, chiefly to keep the buried wires within the right-of-way. The length of the buried wires vary from 200 ft out from the towers to the whole length of the span, that is, connection from tower to tower. The tower-footing resistance in general should not exceed 10 to 20 ohms.

Cost of Transmission Lines

The cost of transmission lines does not follow any recognizable formula. Two identical lines built in the same general locality by different organizations may vary widely in cost. The items that go into cost are conductors, ground wires, hardware and insulators, supporting structures (steel towers or wood poles), and right-of-way. To the cost of these must be added the cost of surveying, access roads, camps, engineering supervision, erection, and interest during construction.

Where it is necessary to make a rough estimate of the cost of a transmission line, the following procedure has been found to give results which do not differ too widely from the few published data.

1. Determine the weight of the conductors and ground wires. Practically all ground wires are 7/16 or 8/16-in. extra high strength steel weighing 2100 and 2700 lb per mile respectively.

2. Multiply item 1 by 2.5 to 5 to obtain the weight of the towers, the smaller figure being for the most economical span in flat country and the higher figure for difficult country requiring extra heavy towers.

3. For the cost of conductor use 6 to 8 cents per pound above the base price of copper or aluminum. The ground wire costs about 18 cents a pound. For the towers use three times the base price of steel.

4. Add the cost of insulators and hardware.

5. Multiply items 1 and 2 by 5 to 10 cents to obtain erection costs. This figure may go higher in remote rough country.

6. To the total cost so obtained, there should be added 20% to cover contractors' fees, engineering supervision, and interest during construction.

As an illustration consider a 230-kv, single-circuit, steel tower line with 954,000 ACSR conductors, two 7/16-in. ground wires, 1000-ft span.

1. Weight of conductors   3 × 6480 = 19,440 lb
   Weight of ground wires   2 × 2100 = 4,200 lb
                                              Total   23,640 lb

2. Weight of towers   2.5 × 23640 = 59,000 lb
   Each tower 59,000/5.3 = 11,000 lb
                             Total weight per mile   82,640 lb

3. Costs:
   Conductors   19,440 × 0.26   = $ 5,054
   Ground wires   4,200 × 0.18   =   756
   Towers   59,000 × 0.11   =   6,490
                                                        $12,300

4. Insulators and Hardware
   5.3 towers = 16 sets $ 1,920
                    Total material per mile $14,220

5. Erection
   82,640 lb at 8 cents $ 6,650
                                                        $20,870

6. Contractors' fees, etc. $ 4,130
                  Total cost of line per mile $25,000

To this must be added the cost of right-of-way. Line surveys cost $100 to $200 a mile in ordinary open country, and clearing the right-of-way costs $600 to $1200 a mile per 100-ft width of right-of-way. Right-of-way easements vary from $300 to $1200. In urban districts right-of-way costs may become prohibitive and make it necessary to go to high-voltage cables.

A more accurate but still approximate method of obtaining the weight of steel towers is given in a paper presented by W. S. Peterson of the Department of Water and Power, City of Los Angeles, at the Paris meeting of CIGRE in 1950. Mr. Peterson found that the following formula could be used with acceptable estimating accuracy for the entire range of high-voltage transmission lines:

$$W = CH\sqrt[3]{P^2}$$

where W = weight of tower structure above ground in pounds,
           C = constant 0.25 to 0.35 for lines 110 kv to 350 kv, depending on variations in tower form,
           H = equivalent height of tower in feet; that is, height above ground at which the resultant pull produces the total moment on tower,
           P = single resultant combined pull in pounds under maximum loading conditions.

Principles of Electric Utility Engineering

The following example of the application of this formula to a 230-kv line is taken from Mr. Peterson's paper:

Conductor, circular mils	954,000
Material	ACSR
Diameter, inches	1,196
Weight, pounds per foot	1,227
Ultimate strength, pounds	34,200
Design tension, pounds	10,250

Span, feet	1,000	1,200
Tower height, feet	91	107
Equivalent height, feet	79	94
Transverse moment (in 1000 ft-lb)	475	674
Longitudinal moment (in 1000 ft-lb)	523	690
Resultant moment (in 1000 ft-lb)	706	964
Tower weight, pounds	10,240	13,330

The tower weights given were obtained by using a constant $C = 0.3$. For example, using the 1000-foot span the resultant pull $P = \dfrac{706,000}{79}$ = 9000 lb. Therefore, $\sqrt[3]{9000^2} = 430$ and $W = 0.3 \times 79 \times 430 = 10,240$ lb.

Automatic Circuit Reclosing

A great saving in transmission-line investment has been achieved in recent years by automatic circuit reclosing. Over 80% of all faults on transmission circuits are single-phase to ground and due to flashovers. If, therefore, the circuit can be opened only long enough to extinguish the arc, and then reclosed before synchronism is lost, the resulting saving is almost the equivalent of a second circuit. The reclosing time of breakers is defined as the time from the energizing of the trip coil until the contacts reclose. Breakers are designed for 30 and 20 cycles reclosing time, these speeds being obtained by energizing the closing magnet before the breaker is fully open.

A certain time during which the circuit is completely de-energized is necessary to extinguish the arc due to the flashover. This will be of the order of 5 to 12 cycles. The breakers at the two ends of the line should, therefore, open simultaneously if maximum benefit is to be obtained from automatic reclosure. Successful application to any particular line requires that the transient stability limit of the line be studied to determine whether it can stand an outage for the time of operation of the relays and circuit breakers.

Higher Voltages

The discussion so far has been restricted to transmission lines of 115 to 230 kv for the reason that, with the exception of the 287-kv line between Hoover Dam and Los Angeles, all transmissions are in this range. It has been evident for some time, however, that the limit would have to be materially increased to meet the tremendous growth in power demand on some of the largest systems. Consequently, there are under construction a transmission of 315 kv on the systems of the American Gas and Electric Company and one of 345 kv on the system of the Bonneville Administration, and many other such lines will no doubt follow. The transmission distances are not necessarily long, but the great capacity of modern generating stations requires the carrying of ever larger blocks of power to load centers. To do this at lower voltages would entail a multiplicity of circuits with prohibitive right-of-way costs.

The problems presented by these lines of very high voltage are economical and operational. The cost of equipment unfortunately does not increase in a straight-line relationship with voltage but a good deal faster. The power transmitted over a single circuit is of necessity very great—it may be as much as 400,000 kw—and the system in case of accident must be able to drop such a circuit without falling apart. The technical problems are well understood and present no particular difficulty. Corona losses and radio interference have been thoroughly investigated here and abroad, and the minimum permissible conductor diameter has been established. The use of "bundled conductors," that is, two or three conductors per phase separated by a few inches has begun abroad but not as yet in this country. The advantages of bundled conductors are that for a given corona limitation smaller conductors easier to handle, can be used, and the reactance is reduced by some 25%. The disadvantages are mechanical complication and possible trouble due to heavy icing in winter.

The insulation of transformers and circuit breakers for these higher voltages presents no new problem. A favorable factor in this regard is that experience with 230-kv systems indicates that the industry has been conservative and the basic insulation levels for very high voltage systems need not be so high as originally anticipated. The cost of transformers for these voltages goes up almost directly with the amount of insulation built into them so that reduction in BIL has economic importance. It is felt that with the transformer effectively grounded and protected the BIL can be given the following values:

| System kv | 230 | 287 | 330 | 380 | 440 |
| BIL kv | 825 | 1050 | 1175 | 1350 | 1550 |

They are based on the performance of lightning arresters under a reasonable combination of probabilities. With a very heavy surge, high rate of rise, unfavorable arrester design tolerance, and poor ground conditions, the margin of safety would be insufficient and damage might result.

Summary

It is seen that the problem of transmission economics involves many factors, which, although interrelated, cannot be reduced to a single formula. It is possible, however, to set down a few general rules:

1. Where a choice of two or more voltages exists within the bounds of reasonable economics, the highest should be chosen. As few different voltages as possible should be used in any one system.

2. The greatest amount of power that stability requirements will permit should be transmitted over each circuit. This usually means operating with a reactance angle between the two ends of the transmission of about 37 degrees. To state it differently, the IX voltage of the line is about 60% of the line-to-neutral voltage.

3. The line insulation level should be selected to keep lightning outages down to two per hundred miles per year. This requirement may mean the use of counterpoise wires in some areas.

4. Insulation coordination is an important factor in keeping down investment. Advantage should be taken of lightning arrester performance to utilize equipment of the lowest basic insulation level permissible in the prevailing circumstances.

5. Ground wires should be located above the conductors in such a manner that the plane through the ground wire and outside conductor makes an angle with the vertical not exceeding 30 degrees.

6. The use of two conductors per phase is worthy of more investigation. Two conductors separated by 6 in. have 20% less reactance than a single conductor of the same total cross-section.

7. The supporting structures of the transmission line are a major item of cost. The possibility of reducing this cost by the use of flexible or semi-flexible towers, towers with tubular members, etc., is worth investigating.

CHAPTER 11

Power-System Stability

The American Standards Association defines power-system stability as "the ability of a system, when disturbed, to develop restoring forces among its elements equal to or greater than the disturbing forces so as to restore a state of equilibrium." The "elements" are the synchronous generators, condensers, and motors. They are tied together electrically by means of transmission and distribution circuits, and synchronism must be maintained between them in steady-state operation and during disturbances. Two types of stability are recognized:

1. Steady-state stability, defined as a condition which exists in a power system if it operates with stability when there is no aperiodic disturbance in the system.

2. Transient stability, defined as a condition which exists in a power system if, after an aperiodic disturbance has taken place, the system regains steady-state stability.

Steady-state and transient power limits are the maximum power flow possible through a point in the system under consideration when it is operating under these conditions.

It is obvious that a knowledge of the factors that enter into the problem of stability is of prime importance to utility engineers. It is also of importance to the manufacturers of equipment, because where stability is likely to be a consideration, special features have to be included in the design which affect cost and performance. Stability is an electromechanical problem, with the inertia of the rotating parts playing as important a part as the electrical characteristics of the machines and intermediate ties.

Power-Angle Diagram

Consider a system transmitting power from a generator G to a synchronous motor M over two lines, each with reactance X, as in

198 Principles of Electric Utility Engineering

Figure 11-1. Under normal conditions the internal emf of the motor lags its terminal voltage by the angle θ_3, and the internal voltage of the generator leads its terminal voltage by the angle θ_1. The total

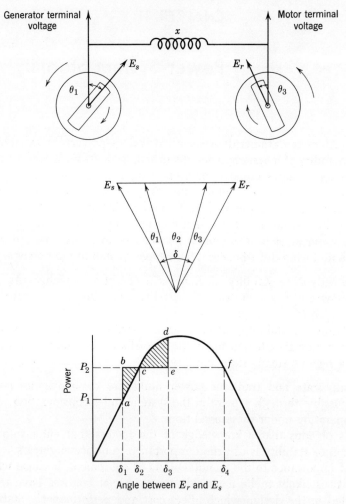

Figure 11-1. Power-angle diagram.

angle between the two emf's is $\delta = \theta_1 + \theta_2 + \theta_3$, where θ_2 is the line angle between the two terminal voltages. If load is added to the motor shaft the rotor falls back, and the angle θ_3 is increased. This change calls for more output from the generator, whose rotor moves forward to increase the angle θ_1. The total angle δ increases cor-

Power-System Stability

respondingly. The power transmitted from generator G to motor M is

$$P = \frac{E_s E_r}{X} \sin \delta$$

where P = three-phase power in watts,
E_s = generator internal voltage,
E_r = motor internal voltage, both line-to-line,
X = reactance in ohms between the internal voltages,
δ = angle by which the generator internal voltage leads the motor internal voltage.

For steady-state conditions, $E_s E_r/X$ is constant, so that if power is plotted against the angle δ the shape of the curve is a sine. The theoretical limit of stability is at 90 degrees. At all angles less than 90 degrees the system is stable (if there is no disturbance), and at all angles greater than 90 degrees it is unstable.

Assume that the system is operating with load P_1 and angle δ_1 and that the load is suddenly increased to P_2. Because of the inertia of the rotating machines, the internal voltages cannot immediately swing to the angle δ_2 corresponding to load P_2. Instead the initial differences are used up in accelerating the generator and decelerating the motor. This change increases the angle, but when the value δ_2 is reached the generator is traveling above synchronous speed and the motor below synchronous speed. The angle is therefore overshot and may reach δ_3. The area abc represents the transient energy causing the overshoot. The area ced represents the energy available for restoring equilibrium. If the latter area is greater than the former, ced greater than abc, stability is restored. If it is smaller, the system pulls out of synchronism. The area $cdfc$ is the maximum available for restoration of stability. If the system swings beyond δ_4, the system cannot recover.

Instability due to a tie-line disturbance, for example, a pole line knocked down, can also be illustrated by the power-angle diagram. Assume that the load on a system is P_1 at angle δ_1 and that one of the tie lines in Figure 11-2 is short-circuited. Under this condition the power transferable from G to M is reduced, but not necessarily to zero. This situation calls for three diagrams: I represents the initial condition with the two lines in circuit; II represents the period during which the fault is on the system; III represents conditions after the breakers have removed the faulted line, increasing the tie reactance from $X/2$ to X. Here again if the area A is greater than B, synchronism is lost. Conversely, if B is greater than A, synchronism is restored. It should be noted that the voltages are internal emf's and that the machine reactances are as much a part of the "through reactance" as the transformers and transmission lines.

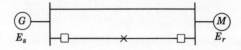

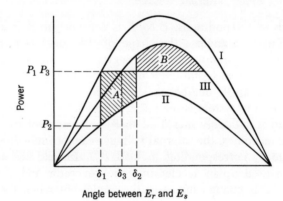

Angle between E_r and E_s

Figure 11-2. Power-angle diagram for fault on one line.

Stability Calculations

The foregoing discussion gives a simple picture of a very complex problem. It ignores many factors which in an actual calculation must be taken into account. A complete stability study makes almost mandatory the use of a network analyzer. The principal factors which complicate the study are the following:

1. Power systems today invariably have more than two termini. It is sometimes possible, but only rarely, to simplify a system to an equivalent two-machine problem. There are therefore a multiplicity of lines and machines to deal with.

2. The tie lines have resistance as well as reactance. The resistance creates loss so that the sending-end and receiving-end maximum powers will occur at different angular displacements. The lines also have capacitance which modifies the reactance X.

3. An obvious method of increasing the stability limit is to reduce

Power-System Stability

the "through reactance." Reducing the spacing between conductors reduces the reactance of the circuits, but the spacing is usually controlled by other factors. Not much can be done to reduce the transformer reactances appreciably, but the method of bussing them (that is, whether they are bussed on the high-voltage side or on the low-voltage side, or both) makes quite a difference. Busses at intermediate points along the line also reduce the "through reactance" by removing shorter sections of lines in case of faults. It is impossible to generalize on the relative merits of high- and low-voltage bussing arrangements. A low "through reactance" is useful at the time the fault is isolated as it invokes less change in system reactance. On the other hand, during the fault the shock to the system is more severe. Each case must therefore be studied for various bussing arrangements, as can be readily done on the calculating board. Other methods of reducing series reactance are by means of "bundled conductors" and series capacitors.

4. Three characteristics of synchronous machines are of importance in the stability problem: transient reactance, inertia, and speed of excitation build-up. Reducing the machine reactance is equivalent to reducing the system through reactance; however, a decrease in transient reactance of one third, or, say, from 30 to 20%, improves the stability limit 5 to 10% in a typical system at a cost increase in the generators of some 30 to 40%. In most cases, therefore, the decrease is not justified economically.

The inertia of the generators affects the period of system oscillation, or the time required to reach the point beyond which recovery would be impossible. The greater the inertia, the lower the acceleration factor. In a few cases where calculations have indicated that the system would operate close to the stability limits, generators of higher than normal inertia have been installed. But here again the gain is relatively small compared to the increased cost of the machines.

Quick-response excitation was one of the early remedies used to improve stability and is now standard practice. It consists of building up the field of the generator exciter immediately on the occurrence of a fault by means of quick-acting voltage regulators so as to maintain (or if possible increase) the machine flux against the demagnetizing action of the fault current.

5. The type and duration of the fault vary. In general, no attempt is made to keep a system stable against three-phase faults. The usual fault used as a basis of system design from the stability point of view is a double line-to-ground fault, but sometimes a single line-to-ground fault is considered. The great majority of system faults are single-

202 Principles of Electric Utility Engineering

phase to ground. The duration of the fault is very important. If the power that can be transmitted over a typical double circuit transmission is 100% for instantaneous tripping, a delay of 10 cycles (60-cycle base) may reduce the stability limit by 25% and a delay of 20 cycles by 50%. This fact explains all the development work that has been done on high-speed breakers and relays.

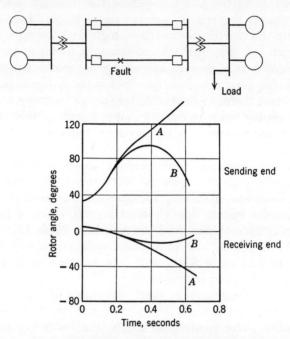

Figure 11-3. Angle-time curve. A. Fault isolated in 12 cycles (instability). B. Fault isolated in 10 cycles (stable).

Figure 11-3 shows the results of a study made to determine the performance of an assumed hydro station in case of a fault on a transmission line connecting it to the load center. It indicates that the fault must be cleared in 0.17 sec or better if synchronism is to be maintained.

Network Analyzer

The network analyzer is a board on which all the elements of a power system can be reproduced in miniature. A generator electrically is essentially two voltage vectors displaced by some angle. It can therefore be represented by a phase-shifting regulator. The circuit elements can be represented by adjustable resistors, reactors,

Power-System Stability

and capacitors. Mutual induction can be represented by 1 to 1 transformers. Jumpers interconnect the various elements to simulate exactly the system being studied. A relatively high frequency supply, 440 cycles or more, is used to keep down the size of the elements. Circuit selector and metering controls are provided for direct readings of either scalar or vector values of current and voltage, watts, vars, and phase angles.

Such a board can be used for the study of stability conditions, that is, the power limits of a transmission system; the study of short-circuit conditions for the application of breakers and relays; the study of voltage regulation, load control, current distribution, etc.

Power Circle Diagram

In a paper before the AIEE in 1904, B. G. Lamme advocated the use of synchronous motors for regulating the voltage and increasing the output of transmission lines, and the theoretical work that followed this suggestion led to the development of the "power circle diagram" for lines with voltages $E_s E_r$ held constant at either end. The circle diagram has lost much of its usefulness with the development of network analyzers, but it is still a convenient means of representing the performance of a transmission line. If capacitance of the line is disregarded, as it can be for lines 40 to 50 miles in length, the construction of the circle diagram is quite simple. On cross-section paper the abscissa represent kilowatts and the ordinates, kva, positive lagging and negative leading.

If E_r = the voltage at the receiving end,
E_s = the voltage at the sending end,
R = resistance of one conductor,
X = reactance of one conductor,

then the center of the receiving end circle is $(a \times b)$, where

$$a = -\frac{E_r^2 R}{R^2 + X^2} \qquad b = +\frac{E_r^2 X}{R^2 + X^2}$$

The center of the sending end circle is $(c \times d)$, where

$$c = +\frac{E_s^2 R}{R^2 + X^2} \qquad d = +\frac{E_s^2 X}{R^2 + X^2}$$

The radius of both circles is $E_s E_r / \sqrt{R^2 + X^2}$.

With these values a diagram is drawn as in Figure 11-4. It shows that at no load a lagging kva (under-excited condenser) of value OA

204　Principles of Electric Utility Engineering

is required to maintain the voltages E_sE_r. If the load transmitted over the line is raised to OB kilowatts, α represents the angle between the voltages E_s and E_r. The same angle appears in both circles. The distance BD represents the leading kva which must be provided at the receiving end if the voltages are to be maintained, assuming unity power-factor load. If the line OF represents the power-factor

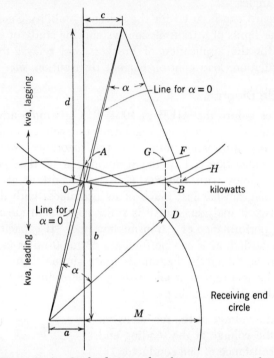

Figure 11-4. Circle diagram for short transmission line.

of the load, additional leading kva (over-excited) of value BG is required to maintain voltages E_sE_r. The distance BH represents the kilowatt loss in the line for load OB. The line M represents the theoretical maximum kilowatts that can be transmitted over the line.

The foregoing simple assumptions make the principle of the circle diagram easily understandable. Stability of systems, as has been pointed out, depends on conditions between the internal voltages of the synchronous machines at the two ends of the line; therefore, the impedance of the machines and transformers must be included. Moreover, in most cases it is not permissible to disregard capacitance. Methods of constructing the diagram for these more complicated conditions are found in textbooks on transmission-line design.

Series Capacitors

As previously stated, a rather obvious method of improving system stability is to reduce the through reactance, which can be done with series capacitors. Reactive power results from the collapse of electromagnetic and electrostatic fields. It flows through the system in the same manner as active power and causes both an energy loss and a voltage drop even though it provides no useful work. Shunt capacitors, that is, capacitors in parallel with the load, correct the component of current due to inductive reactance. Series capacitors compensate for the reactive voltage drop in the circuit. Shunt capacitors are used widely for power-factor correction. Series capacitors are used in low-voltage heavy-current applications, such as furnaces and welders, to compensate for the voltage drop in the conductors.

In recent years series capacitors have been used in high-voltage transmission lines to compensate for the drop due to inductive reactance and to increase the power transmitted. If the reactive drop is completely compensated for, the resulting drop is that due to resistance. Complete compensation is generally not desirable for stability reasons. During line faults, the fault current produces an excessive voltage across the capacitor which makes it essential that it be taken out of service very quickly. But taking the capacitor out of service is equivalent to adding reactance to the circuit, which is the worst thing that can be done at a time of fault from the viewpoint of maintaining stability.

The first series capacitor in a transmission circuit was probably one installed in Japan in 1945, and one or two more installations followed abroad. These capacitors in case of a line fault are short-circuited and reinserted in 30 cycles. They would not be considered satisfactory in this country because with any worth-while compensation the two ends of the circuit would be out of synchronism in 30 cycles. The series capacitors installed in this country are not short-circuited in case of a line fault, but rather a resistor is placed in parallel with the capacitor, having a value such that with the stated fault current the voltage across the capacitor is limited to a safe value. The resistor is inserted by the breakdown of a gap which acts as a switch. The resistor is therefore in circuit from the moment the voltage reaches the breakdown of the gap until the gap arc is terminated. The gap is designed in such a manner that the arc is extinguished at each half cycle. Since the resistor is reconnected at each half cycle if the line fault is still present, it follows that once the line current is normal the resistor is not reconnected, and the series capacitor is then 100% effective. Means are provided to short out the series capacitor in emergency conditions, such as equipment failure. A typical circuit arrangement of gaps and resistors, and the gap itself, are shown in Figure 11-5. The resistors limit the oscillatory discharge current so that the gap can break down and restrike repetitively without damaging the capacitor. The gap consists of two graphite electrodes arranged so that compressed air can be forced through the arc and exhausted to the atmosphere.

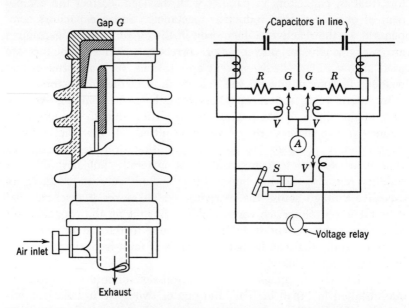

Figure 11-5. Typical protective scheme and gap for series capacitor. R. Resistor. G. Gap. V. Valve. A. Air reservoir. S. Short-circuiting switch.

The manner in which this scheme affects the stability of a system is illustrated in Figure 11-6 for the case of two parallel lines with an intermediate sectionalizing station with series capacitor. The figure shows the power-angle diagrams for the system operating normally, for the system with an assumed fault to ground, for the system with the faulty line section removed but the capacitor still short-circuited, and for the system with the faulty line out and the capacitor reinserted. The initial operating condition is unit power and an angle of 30 degrees. The fault reduces the power flow from a to b at the first instant of time, and the angular shift moves the operating point to c. At this point the faulted line section is removed and the operating point moves up to f with the capacitor reinserted in the circuit. The

Power-System Stability

restoring forces are shown by the area *dfikg*. Without the capacitor the restoring forces would be as indicated by the area *dheg*. Thus the series capacitor, if properly applied, greatly increases the restoring forces and the system stability during faults, provided that it does not have to be taken out of circuit to protect it against destruction.

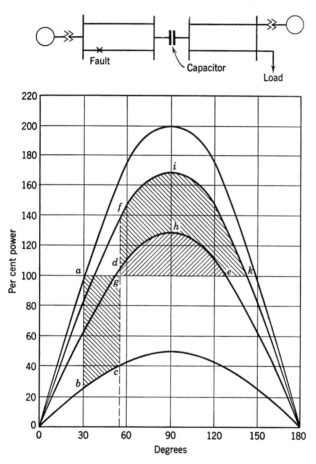

Figure 11-6. Power-angle diagram of transmission system with intermediate series capacitor.

Unity Power-Factor Operation

As mentioned above, shunt capacitors are used extensively to compensate for the lagging power factor of induction motors. The capacitors are usually mounted on poles or otherwise scattered around the distribution system, and for the most part there is no way of discon-

208   Principles of Electric Utility Engineering

necting them from the system with variations in load. The result is that at night and over the week-ends when the induction motors are idle the system power factor may approach or surpass unity. In these circumstances the generators are underexcited, and the margin of stability is reduced. If at such times load is suddenly thrown on a generator because of relay operation elsewhere on the system, the generator may be unable to pick it up and the system loses synchronism. The problem is essentially one of steady-state stability but is so complicated that an analytical solution is practically impossible. Fortunately modern voltage regulating systems build up excitation quickly enough to keep the generators in synchronism in the majority of cases of trouble. Nevertheless, operation at high power factor is a problem which in all power systems must be given careful consideration.

System Loading Schedules

Closely tied in with the problem of stability is the matter of system operation and the allocation of generator capacity in relation to system load and circuit conditions. The choice of the stations where load increments should be placed is frequently determined by maximum economy requirements, which may not be the best choice from considerations of system stability. Adequate spinning reserve capacity in the form of spare generators or reliable interconnections must be available in each load area to insure stability in the event of the loss of the largest unit in the area. The problem of a proper balance between loading for economy and generating for stability is an operating compromise, and from time to time cases occur where synchronism is lost because of system faults.

In all power systems one or more stations or generating units (the most efficient) operate at constant output as base-load units at the point of best efficiency, but this condition still leaves a number of units which must put out energy to conform with the load curve. The apportioning of the load among these units is known as "incremental loading." The general theory of incremental loading is treated in a thorough manner in Steinberg and Smith's book *Economy Loading of Power Plants*, to which the reader is referred. It is shown that maximum efficiency, that is, minimum combined input for a given combined output, is obtained when machines are operating at outputs which correspond to the same "incremental rate value." By incremental rate value is meant the rate of change of input with respect to output. For instance, if the increase of output from 4000 to 5000 kw in a turbine generator involves an increase in steam equiva-

Power-System Stability

lent (input) from 19,000 to 21,000 kw, then the incremental rate $= \dfrac{\text{increment input}}{\text{increment output}} = \dfrac{2}{1} = 2.$ Another way of stating this axiom is that if the outputs of two machines change at the same rate as their inputs, the machines have the same efficiency.

Incremental rate curves can be determined with sufficient accuracy from the input-output curves of the various pieces of equipment by dividing the curves into suitably small intervals and dividing the change in input by the change in output.

In order to operate at maximum economy the efficiency of all equipment from the coal pile to the point of added load, including transmission losses, must be taken into account. In a combined steam and water power system the calculations are likely to become quite involved because of the variability of the pondage and the consequent frequent changes that have to be made in the supply circuits.

Frequency Control

Frequency is maintained by the utilities at 60 cycles not, as is widely assumed, to keep electric clocks right, but for economic reasons. If the frequency is uncontrolled, interchanges of energy of considerable magnitude between interconnected systems result. No utility "sells time," and electric clocks may vary at any instant from correct time by several seconds.

Maintaining frequency is the balancing of power supply with the prevailing consumer demand. This is accomplished through changing the setting of the governor, manually or automatically. Except for certain special applications all turbines have a "drooping characteristic," that is, for a given valve opening the speed drops with load. Restoration to normal speed is obtained through raising the speed-drop curve parallel to itself. Automatic control is merely doing by some suitable mechanism what an operator would do by hand. It does not supplant the governor. The inherent speed droop of the governor is the amount of permanent speed droop the governor allows between no load and full load. This droop is necessary to obtain stable load division between turbines. In large units it is usually 4%. A good governor properly adjusted has a sensitivity of 0.05%; that is, this change in speed is required to make the governor respond. Every governor also has a "dead band," that is, a small range of speed within which the governor will not respond to make a correction. This is necessary to provide stability to the valve mechanism. The dead band in large units is 0.06%; in small units, 0.1%.

It is not possible by means of the governor alone to control both speed (frequency) and load. If we design the governor to maintain constant speed (so-called isochronous governors), there is no control over the load. If we design the governor with a droop, the frequency varies with load. Operating experience has shown that it is more practical to design the governor for load control, and use an external means to obtain constant frequency, that is, to compensate for the change in speed from no load to full load.

There are today essentially two methods of frequency control. In one, a station (usually the largest) is assigned to correct the frequency for the whole system. In the other, each load area is expected to take care of its own control.

Where a load is carried by a single generating station, frequency presents no problem. The difficulty comes when two or more stations are operating in parallel, as they are in all utilities today. The governors in the various stations have different sensitivities, dead bands, and droops, and, when an increment of load is added somewhere on the system, the turbine with the quickest "response" picks it up, whether or not it is desirable that it do so. If two utilities of different ownership are tied together, it may well happen that changes of load on one system are corrected by the other system.

To prevent this sort of random accommodation of load changes, a system of tie-line control was developed whereby the power flow over the tie line was limited by a watt relay. In recent years the two functions of tie line and frequency control have been combined in what is known as "tie-line bias frequency control." In this scheme the system is divided into control areas in which some station is assigned to maintain frequency. If there is a sudden increase in load within such an area, the first accommodation may come over the tie line because of relative governor characteristics, but the tie-line control relay, sensing an overload, then acts to force the turbine in the area to take up load until the tie-line power flow is back to normal. The bias control does not necessarily bring the tie-line power flow back to its original value. It can be set to permit an increase to allow all areas to share in the general increase of system load to any desired extent. This is equivalent to moving the droop curve of the turbine governors up more than is required merely to correct frequency within the area. The tie-line load control should act on the same station as the frequency control, since otherwise different droops in the various stations may cause instability.

There are two fundamental principles forming the basis of frequency control, one being response to accumulated time error, the other response to deviation from standard frequency. The first is exemplified by the Warren pendulum clock controller, the second by the Leeds and Northrup tuned-bridge controller.

The first consists of a fine clock mechanism with a pendulum that makes a complete swing in two seconds. The pendulum drives the clock hand through a differential in one direction. A synchronous motor drives the clock hand through the same differential in the opposite direction. If the frequency varies from normal, the clock hand shows the time error. At the same time a cam mechanism makes contact in a circuit that sends impulses to the synchronizing motor on the turbine governor, the contacts being held closed for a length of time that is proportional to the integrated time error during each two-second interval.

The Leeds and Northrup control uses a tuned-frequency bridge. If the frequency deviates from normal, contacts are closed and impulses sent to the turbine synchronizing motor. The period of the contact is proportional to the deviation of the frequency from normal, this proportionality being achieved by the well-known Leeds and Northrup Micromax tilting-rocker mechanism used in most of their instruments.

In the early days of frequency control, it was thought necessary to bring the frequency back to normal, after a deviation, as quickly as possible. This action usually resulted in serious power swings over tie lines, and more recently it has become the practice to make corrections more slowly. If time wanders from correct time by some seconds during the day, a correction is made during light load period at night to bring it back to normal.

During the war a shortage of power developed in some localities, particularly where prolonged droughts cut down the hydro capacity. To help out the situation, experiments were made with lowering the voltage and frequency in New Jersey, Pennsylvania, the Middle West, and California. A reduction in frequency means a reduction in speed of the rotating equipment, and at slower speeds less work is done. The results of the tests were remarkably similar in the different areas, indicating that the composition of the load is much the same in the different areas. It was found that with constant voltage a 1% reduction in frequency results in a 1.9% reduction in load. A 2.5% reduction in frequency reduces the load 4.7%. A 1% drop in voltage adds a further 1% reduction in load.

As a result of these tests, it was felt that in an emergency it should be possible to drop the voltage 5% and the frequency 1.5% and achieve a reduction in load of 9%. The public should be notified beforehand of the drop in frequency because of its effect on electric clocks.

CHAPTER 12

Power Distribution

Most of the customers of the public utilities can be classified as industrial, commercial, or residential. Most industrial and all commercial customers are supplied at distribution voltages. The voltage for residential service in this country is uniformly 120 volts. Abroad it is mostly 200 or 230 volts. This fact makes for considerable differences in the method of distribution. In this country the voltage is reduced in a distribution substation to a value of 2400 to 13,800 volts, and a second reduction is made to utilization voltage (120 volts) by means of a large number of small transformers mounted on poles, or located in vaults, known as distribution transformers. Abroad, the subtransmission voltage is brought into a kiosk or vault and reduced to 230 volts, and a relatively large area is covered at 230 volts. Transformers of less than 500 kva are seldom used.

Distribution is defined as that part of the utility system between the bulk power source and the customer's service switch. The bulk power source may be a generating station or a power substation. On this basis the utility investment in distribution is about half of the total investment in the system. It is somewhat less than half where the distribution is mostly open-wire overhead and somewhat more where a large part is in underground cables.

The distribution system is divided into three parts: (1) the subtransmission circuits and distribution substations, (2) the primary distribution circuits, and (3) the distribution transformers and secondary circuits.

Subtransmission System

The subtransmission circuits are those that bring power from the bulk supply stations to the distribution substations. The voltage of these circuits varies from 13,200 to 69,000 volts, with the majority at 34,500 and 46,000 volts. The circuits may be overhead open-wire construction on wood poles or cable in ducts underground.

Power Distribution

The principal circuit arrangements are radial, loop, and grid, with combinations and modifications. The choice depends on the transmission system used and the nature of the territory served. Where the subtransmission circuits are used to interconnect bulk supply stations as well as to supply the distribution substations, an electrically rigid system is necessary for stability reasons.

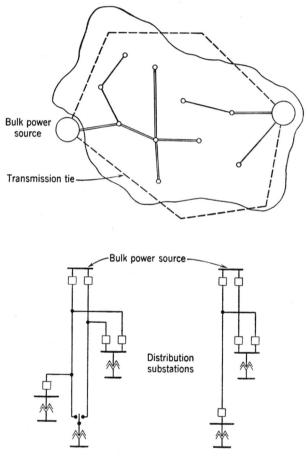

Figure 12-1. Radial-type subtransmission circuits.

In the radial system (Figure 12-1), as the name indicates, the circuits to the distribution substations radiate from the bulk-power stations. Almost invariably there are two circuits in order to ensure supply to the substation in case of a line fault. Where possible, the two circuits are not mounted on the same poles.

In the loop arrangement (Figure 12-2) a single circuit starting

214 Principles of Electric Utility Engineering

from one of the bulk-power stations passes through a number of substations and returns to the same station. The advantage of this scheme is that every substation is supplied by circuits on different rights-of-way. This fact permits overhead open-wire construction where otherwise cable might have to be used.

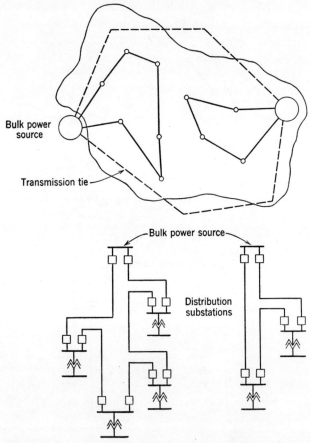

Figure 12-2. Loop arrangement of subtransmission circuits.

In the grid system (Figure 12-3) no attempt is made to restrict the supply of any substation to one bulk-power station. The stations are tied together as in the loop system, but the subtransmission circuits may start from one bulk-power station and finish up at another. The difficulties with this system lie in the control of power flow and relaying. The grid is nevertheless the most extensively used system, though the trend is toward simpler ones.

Power Distribution 215

The subtransmission lines are for the most part of wood pole construction with pin-type and post-type insulators on wood cross-arms. The cross-arms vary in length from 6 to 10 ft. Where a ground wire is considered unnecessary, a common arrangement is a single cross-

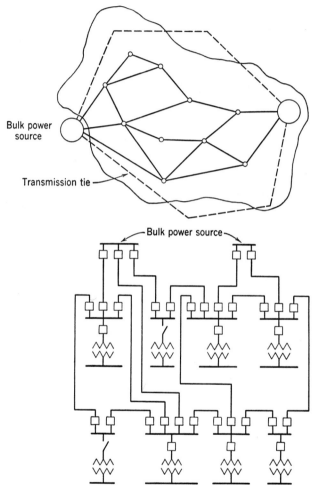

Figure 12-3. Grid system of subtransmission.

arm with the middle phase insulator on top of the pole. However, since the isokeraunic level in most of the country is 30 or more and justifies a ground wire, two cross-arms are generally used with the ground wire at the top of the pole. The vertical clearance from ground wire to the upper cross arm and between cross arms is 4 to 6 ft. In a territory with 30 to 40 lightning storm days per year, 40 to 50 trip-outs due to lightning may be expected per 100 miles per year if no ground wire is used. With a ground wire and proper utili-

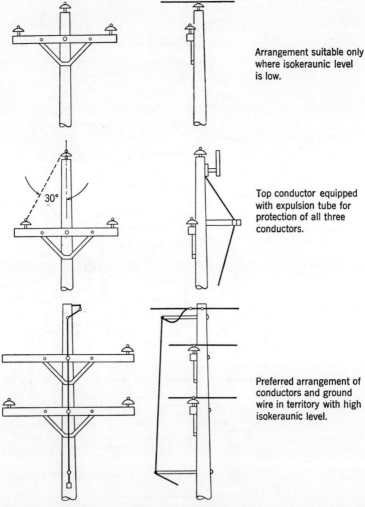

Figure 12-4. Typical arrangement of conductors for subtransmission circuits.

zation of the insulating value of wood, practical immunity from lightning outages is possible. Wood has an impulse strength of 50 to 100 kv per ft, depending on the moisture content, so that even a foot of pole adds percentagewise considerable strength to pin-type insulators. In order to take advantage of this added strength, it becomes

Power Distribution 217

necessary to keep the down-lead from the ground wire away from the pole for some distance. Special hardware for this purpose is obtainable from several manufacturers. Commonly used conductor configurations are shown in Figure 12-4.

Though the trend is toward small factory-built stations of 1500 to 7500 kva spaced more closely together, most substations are still large, 10,000 to 40,000 kva. These large substations usually have a high-voltage bus and two or three banks of single-phase transformers supplying one low-voltage bus. The high-voltage bus is sectionalized,

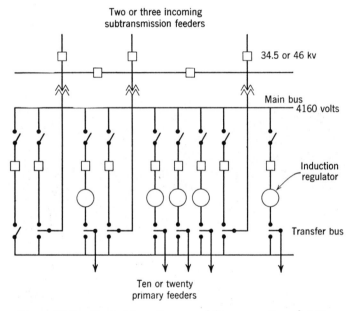

Figure 12-5. Typical bus diagram of large-capacity substation.

the sections being connected by automatic breakers operated normally closed, but arranged to open and isolate a faulted section by differential protection in case of a bus or transformer fault. A large substation of this type has on the low-voltage side a main bus and a transfer bus as shown in Figure 12-5. When maintenance work has to be done on the regulator or breaker of an outgoing feeder, that feeder is temporarily connected to the transfer bus and the equipment isolated from the main bus. There are variations of the scheme shown, such as the use of a high-voltage ring bus or two low-voltage main buses, but Figure 12-5 may be considered representative of general practice.

218 Principles of Electric Utility Engineering

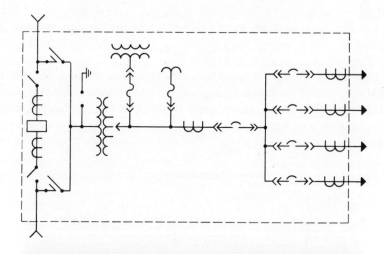

Figure 12-6. Typical small substation supplied from tie-subtransmission circuit, 15 to 69 kv, with automatic sectionalizing by high-voltage breaker. Primary feeder voltage 2.4 to 13.2 kv. (Courtesy Westinghouse Electric Corporation.)

Power Distribution

With the trend to small substations comes also a trend to simpler circuit arrangements, radial and loop systems rather than grid systems. The small substations use three-phase rather than single-phase transformers, and voltage regulation is obtained by devices for tap-changing under load which are built into the transformer. Advantages of the small stations are that they can be factory-built; they can be loaded nearer their capacity; they make for a flexible adaptation to variations in load density; they can be located in residential areas without marring the appearance of the district; they shorten the length of the distribution circuits; and in some cases they can dispense with high-voltage circuit breakers. Another important fact is that in the case of transformer trouble, or at times of necessary maintenance, a mobile substation mounted on a trailer can quickly take over, so that spare equipment is reduced to a minimum. Figure 12-6 shows a typical small substation with its wiring diagram.

Primary System

The primary system includes that part of the distribution system between the substation and the distribution transformers. Four-wire, wye-connected circuits at 2400/4160 volts are the most widely used. Three-phase, four-wire, 7620/13,200-volt circuits are used for rural distribution, and in a few cases for urban and suburban distribution, but the economic advantage of the higher voltage for suburban distribution is still unproved. The circuits from the substations are usually referred to as "mains" and the branches from the mains as "laterals." The neutral is grounded at several locations. Where primary and secondary circuits are on the same poles, a common neutral conductor is used for both systems.

The conductors used for overhead primary circuits are mostly copper of 1/0 and 2/0 cross-section. Larger copper conductors are expensive to install and maintain. With the relative change in the costs of copper and aluminum, the latter is finding ever greater use; and it would appear that most of the primary circuits put up in the future will be of aluminum of 3/0 and 4/0 cross-section.

Underground primary feeders are used in congested and heavy-load areas. They are usually three-conductor cable of 4/0 and 350,000 CM section laid in ducts. Lead-covered cable is the preferred type, but less costly cables are obtainable and are extensively used. In some installations cable is used suspended on poles. The advantages of this method over open-wire construction are improved appearance and lessened voltage drop. The cable is usually single-phase, twisted round and lashed to a copperweld messenger cable which acts as the neutral wire. The cost of such installations is greater than that of open-wire but much less than that of underground construction.

Protection

The problem of distribution protection is complicated because, in case of a fault, it is desirable to remove as little as possible of the faulted circuit so as to interrupt service to as few customers as possible; yet the cost of protection must bear some reasonable relation to the revenue obtainable from the circuit being protected. This means that protective schemes involving circuit breakers and relays cannot for the most part be justified economically.

The primary feeders supply power to the distribution transformers. The faults that can occur therefore are failure of the transformers or of the feeders themselves. The transformers are connected to the feeder through a fuse known as a "primary cutout," which consists of a special alloy wire or strip mounted in a weatherproof porcelain housing suitable for mounting on the cross-arm. The housings are usually arranged to open when the fuse blows and thus provide an indication of operation visible from the sidewalk. It is common practice to fuse 2400-volt transformers (one phase to neutral in 4160-volt systems) on the basis of 1 amp per kva rating with a minimum of 5 amp, for example, a 50-amp fuse for a 50-kva transformer and a 5-amp fuse for a 3-kva transformer.

Where a feeder is of considerable length supplying many transformers, it is undesirable to drop the whole feeder for a fault some distance away, perhaps on a lateral. Therefore, between the breaker at the primary substation and the last distribution transformer the feeder is sectionalized by means of reclosing devices at several locations, usually where laterals are tapped off. The operation of all the fuses and reclosers must be so coordinated that only the fuse or recloser nearest the fault on the supply side locks out if the fault is sustained.

The devices available for this type of protection are single-phase reclosing fuses, single-phase sectionalizers, single-phase automatic reclosers, and three-phase automatic reclosers. The circumstance that makes their use economical is the fact that the great majority of faults can be cleared through opening the circuit for only a few cycles. The single-phase reclosers and sectionalizers were devised to replace the repeater fuses, and, when first cost, installation, and maintenance are taken into account, they probably show greater economy than the fuse. The three-phase recloser is used on three-phase circuits on which there may be motors. On such circuits a single-phase recloser

Power Distribution

on locking out would leave the motors operating single-phase, which is undesirable.

Reclosers are essentially circuit breakers with relatively little rupturing ability, equipped with a mechanism to open and then reclose the circuit automatically. They are usually made to operate four times before locking out. These four openings are adjustable in time to permit coordination with other reclosers or fuses. The sectionalizer is similar except that it has no rupturing ability. It is designed to open the circuit at zero volts, that is, after the circuit has been opened by a recloser or breaker.

Where the rupturing requirement is more than can be handled by a recloser, circuit breakers with automatic reclosing relays become necessary. A typical scheme of protection consists of such a reclosing breaker at the substation and fuses out in the mains and laterals to blow in case of sustained faults. The operation is such that when a fault occurs some distance along the circuit, the substation breaker opens and recloses with no or very little (2 cycles) intentional delay. During the reclosure the trip circuit is transferred from the instantaneous to time-delay relays. If the fault is transient (and most of them are), it will have been cleared by the breaker operation. The reclosing relay recycles and transfers the tripping back to the instantaneous relay, the recycling time being 7 to 10 cycles. If, however, the fault persists when the breaker recloses, the time delay permits the fuse to blow and remove the faulted section. If for any reason the fault is not cleared by the sectionalizing fuse, the breaker is tripped by the time-delay relay and the reclosing mechanism is locked out. Of course, in this event all the customers on that feeder are out of service.

Successful performance of a sectionalizing device, switch or fuse, depends on time selectivity, and the manufacturers have succeeded in making fuse links that have remarkably consistent performance curves. The selection of fuse links follows a calculation of magnitude of expected fault current at the fuse locations and at the ends of the laterals. The links should be so chosen that the minimum blowing time is not less than 0.25 sec so as to give the station breaker time to go through its reclosing cycle.

Secondary System

The secondary system consists of the distribution transformers and wires to the consumers' service entrances. The standard utilization voltage is 120/240 volts, three-wire, single-phase, from which both 120-volt lighting and 240-volt single-phase power connections are

222 Principles of Electric Utility Engineering

made. For power use, 240- and 480-volt, delta, three-phase voltage is provided. For combined light and power, 120- to 208-volt, four-

Figure 12-7. Modern single-phase distribution transformer incorporating lightning and overload protection. (Courtesy of Westinghouse Electric Corporation.)

wire, star voltage is used. The last is common practice in metropolitan secondary network systems.

Distribution transformers for open-wire systems are single-phase, oil-filled, two-winding units suitable for being mounted on poles. It

Power Distribution

is usual practice to provide every transformer with a fuse cutout and a lightning arrester. In recent years the transformer has been equipped with an automatic breaker and lightning arrester to make it completely self-protecting. The circuit breaker will open and disconnect the transformer from the system in case of overloads and secondary circuit faults. Outages due to lightning have been eliminated by the use of arresters connected between the high-voltage leads and the tank. Insulation from low-voltage coil to ground is protected by bushing gaps between the low-voltage bushings and the tank. Gaps can be used for 120 volts because an arc will not be maintained at this low voltage, so that when the lightning surge current has passed the power-follow current will be broken at the first current zero. A high-voltage fusible element renders protection to the high-voltage distribution system in the remote case of a fault within the transformer itself. Such a CSP transformer is shown in Figure 12-7.

Overhead secondary circuits are usually radial, but there are many overhead secondary networks. Fractional horsepower motors and household appliances are supplied from the 120-volt lighting circuits. Ranges, water heaters, and single-phase motors up to 5 hp are supplied from the outside wires of the three-wire system at 240 volts. Most existing residences have only a two-wire, 120-volt supply, but the increased use of household appliances has led to the practice of putting three-wire, 240-volt supply into most new residences.

Studies made to determine the proper size of overhead conductor to give good voltage regulation and minimum cost per kva of load for single-phase 120/240-volt systems indicate that three conductors of 40,000 to 70,000 CM section are the best, that is, AWG No. 4 to No. 2, the latter being the preferred size. The permissible length of three No. 2 mains, on the basis of evenly distributed load and 2.5% voltage drop, is as follows: 7.5 kva per 1000 ft, 750 ft; 15 kva per 1000 ft, 520 ft; 22.5 kva per 1000 ft, 420 ft; and 30 kva per 1000 ft, 350 ft.

Distribution transformers for overhead systems are mostly 10, 15, and 25 kva capacity. They are usually loaded until the peak load reaches 150% of their rating, and are then replaced or additional capacity is provided through putting in intermediate transformers.

Network Distribution

The discussion so far has dealt with "radial systems," that is, systems in which energy is distributed from a source as along the spokes of a wheel. There is a second system of distribution known as a "network," in which the supply lines are all tied together in a mesh or grid with suitable protective means. Most overhead distribution is

224 Principles of Electric Utility Engineering

radial, but most underground distribution is of the network type. Underground distribution costs several times as much as overhead construction and is used where the load density is high—50 to 250 kva per 1000 ft. Since the transformers have to be put in vaults, larger units are used than in overhead distribution and a voltage drop of 4% is allowed. The circuit lengths in this case are: for 50 kva per 1000 ft, 1/0 copper, 350 ft; for 100 kva per 1000 ft, 4/0 copper, 350 ft; and for 200 kva per 1000 ft, 4/0 copper, 200 ft.

Where the load density justifies underground construction, it also justifies a secondary network. In other words, if an underground radial system is used, the reason is probably something other than load density, for example, a desire not to spoil the appearance of a high-class residential development by overhead lines. The underground network system provides the highest class of service; outages on such systems are practically unknown.

The 120/208-volt secondary mains surrounding city blocks are connected together to form a grid which is supplied by a plurality of high-voltage feeders through network transformers. The failure of one feeder does not cause any interruption to service because the load on the grid is supplied over the remaining feeders. When a fault occurs in a high-voltage feeder, the breaker at the supply end of the feeder opens and disconnects it from the distribution station bus. At the same time, all the network transformers connected to that feeder are disconnected to prevent back-feed of power from the grid into the fault. This is done by the network protector, which consists of an air-circuit breaker with a closing and tripping mechanism controlled by a network master and phasing relay. When the network protector is closed, the relay functions to trip it on reversal of power flow. The master relay and the phasing coil act together to close the protector when, and only when, the correct voltage conditions exist across it. The principle is illustrated in Figure 12-8.

The protectors are designed to open on the small amount of power which flows when the feeder is disconnected at the supply end and its transformers are excited from the grid. It is thus possible to isolate any feeder completely through merely opening the station breaker. To put a feeder back in service, it is necessary only to close the station breaker. The protectors take care of the rest.

Three-quarters of the network protectors in use are of the submersible type with ratings from 800 to 3000 amp. Open-type protectors are available and are used in building basements or other places where the probability of submersion is slight. The open-type protectors are enclosed in steel-sheet ventilated housings.

Power Distribution 225

Network transformers differ from ordinary distribution transformers in that they are always three-phase; they have high-voltage disconnecting and grounding switches and manually operated no-load tap changers. For vault installation the transformer has the high-voltage

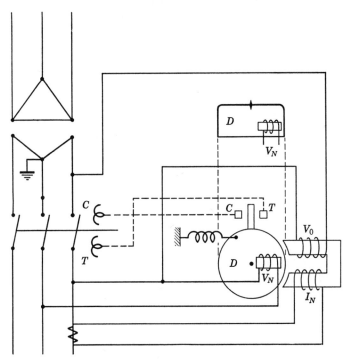

Figure 12-8. Principle of network protector relay.

C = Closing coil and relay closing contact
T = Tripping coil and relay tripping contact
D = Relay drum
V_N = Voltage coil across secondary voltage
I_N = Current coil across secondary current transformer
V_0 = Voltage coil across secondary breaker

V_0 and I_N together form a watt relay to rotate drum and close trip contact T on reversal of power (network to feeder). With voltage restored V_0 and V_N rotate drum to close contact C and the circuit breaker.

switch mounted on one end and the protector mounted on the other, the whole capable of withstanding submersion. The reactance is usually 5%, this being a compromise between the high value (7 to 10%) that would ensure the best load division among the transformers, and the low value (3 to 4%) that would result in better voltage regulation.

The secondary mains usually follow the pattern of the load area, being located in ducts under the sidewalks or in alleys. Service connections are made in manholes, vaults, or junction boxes. An ideal pattern is obtained when the city is laid out in rectangular blocks and the vaults can be located near street corners.

Single-conductor cable is used because of the many connections and taps that are necessary. Because of its lower cost, cable with nonmetallic sheath is largely used today in preference to lead-covered cable. The single-phase conductors are twisted where possible to keep down the reactance drop of the circuit.

The size of conductor is chosen to keep the voltage drop along the mains under normal load conditions to not more than about 2%. The carrying capacity of the secondary mains should be not more than two thirds of the rated capacity of the predominant size of transformer unit, because a part of the maximum load on the transformer comes from the other mains connected to the same junction. Moreover, some margin is required in the transformers to take care of the load when one primary feeder is disconnected. If two parallel mains are located in duct banks, one on either side of the street, and are tied together only at street intersections, each main should have a current-carrying capacity equal to about 38% of the rating of the largest transformer associated with it.

The conductor sizes mostly used are 4/0, 250, 350, and 500 MCM. In place of one 500,000-CM conductor, it is preferable to use two 4/0 or two 250,000-CM cables because they are easier to handle, faults burn clear more readily, and the voltage regulation is improved. At 80% power factor, the improvement in regulation of two 250,000-CM cables over one 500,000-CM cable is 30%.

The secondary network system is made possible by the fact that at 120/208 volts arcs are not sustained. At higher voltages, say, 480 volts, this method of clearing faults is not dependable. Where the current available is insufficient to burn a fault clear, a so-called "limiter" is used. The limiter is a restricted copper section installed in the secondary main at each junction point. The fusing characteristics of the limiter are designed to clear a faulted section of main before the cable insulation is damaged by the heat generated in the cable by the fault current. In some networks limiters are put in all mains on the basis that the saving in damage to cables more than pays for the limiters.

The high-voltage switch mounted on one end of the network transformer may be a two-position grounding switch, a two-position disconnecting switch, or a three-position disconnecting and ground

Power Distribution 227

switch. The two-position grounding switch does not disconnect the transformer from the feeder. It merely grounds the feeder at the transformer terminals. The two-position or three-position disconnect-

Figure 12-9. Three-phase network transformer, submersible type for use in underground vaults. Three-core primary cable entrance and disconnecting switch at left; secondary network protector at right. (Courtesy Westinghouse Electric Corporation.)

ing switch is preferable. In any case the switch is manually operated and is not designed to open current. It is merely a disconnecting device. A typical network transformer is illustrated in Figure 12-9.

Overhead secondary networks follow the same pattern as underground networks. Frequently extensions of the underground network are carried overhead in areas where the expense of the underground construction is not justified. The load density being lower, the transformers and protectors are smaller but operate in the same manner.

Normally all primary feeders and all network transformers are in operation and carry their proportionate share of the load. As loads change, the division of load changes so that equal voltage drop is maintained from adjacent transformers to every point on any interconnecting main. Thus for any given load conditions, the least possible voltage drop to services is obtained. Since there are at least two paths of supply to any load tap, abrupt changes in load, such as the starting of motors, cause less disturbance than in any radial system. Economically the secondary network system can always be justified where high quality of service is required. For equal reliability any other system becomes equally expensive at any load density. For high load density, the network system is the most economical irrespective of quality of service.

Voltage Regulation

It is not possible to maintain on any given circuit a single value of voltage for all locations and hours of the day. Voltage varies, depending on both the distance of the service switch from the supply transformer and the load taken, which varies from morning to night. Voltage is one of the most important parameters in the design of equipment, appliances, and devices, and satisfactory performance is not to be expected with wide variations in the supply voltage. It is these facts that raise the important question of voltage regulation. The definitions of voltage are:

The "nominal voltage" of a circuit or system is the value assigned to it for the purpose of convenient designation.

The "rated voltage" is the voltage value to which operating and performance characteristics of equipment and apparatus are referred.

The "mode voltage" is the value of voltage which occurs most frequently. It is of significance only in evaluating the performance of some types of equipment.

Fortunately, most appliances and devices operate satisfactorily over some range of voltage so that a reasonable tolerance is permissible. The total range between the maximum and minimum voltages

Power Distribution

on a system during the whole course of its operation is designated "voltage spread."

For any given nominal voltage, the total actual operating values in the industry as a whole varies over a considerable range. This range has been divided into three zones: (1) the "favorable zone" containing the greater part of the existing voltages, (2) the "tolerable zone" containing values above and below the favorable zone that for most purposes will still be acceptable, and (3) an "extreme zone" containing voltages on the fringes of the tolerable zone which may or may not be acceptable, depending on the type of application. The extreme limits represent the "voltage spread." It is not thought that any one company operates over the whole spread, but rather that the lowest value of the company operating in the lower zone and the highest value of the company operating in the higher zone constitute the minimum and maximum of the range of "voltage spread."

The standard United States utilization voltage has a nominal value of 120 volts, a favorable zone of 110 to 125 volts, a tolerable zone of 107 to 127 volts, and an extreme zone of unspecified limits. The extreme departures in voltage from normal should be considered as temporary and subject to improvement.

Originally power was sold at 100 volts, but the voltage has been increasing ever since. The lack of uniformity in utilization voltage made difficulties for the designers of appliances, so that in 1949 the Edison Electric Institute and the National Electrical Manufacturers Association adopted 120 volts as standard.[1] Through a comprehensive statement of the problem and through setting up operating limits, it is hoped to halt any further "inching up" of the voltage. Equipment and appliance voltage ratings have been adopted to conform with the 120-volt supply voltage.

It would be uneconomical to design a distribution system with inherent voltage drop from no load to full load and from first customer to last customer within acceptable limits of voltage regulation. Therefore, automatic voltage regulation is always provided. The three methods used are: (1) bus regulation in the substation, (2) individual feeder regulation in the substation, and (3) regulation along the main by regulators mounted on poles.

In designing urban feeders, it is common practice to keep the voltage range between the first customer and the last within 10%. The maximum range will occur when the first customer's load is

[1] "EEI-NEMA Preferred Voltage Ratings for A-C Systems and Equipment," EEI Pub. R.6 NEMA Pub. 117, May, 1949.

light at the time of peak load and the last customer's load is heavy. At peak load on the feeder, the voltage regulator will be in the maximum boost position. If then the first customer has little load, the drop through the transformer and service leads will be small,

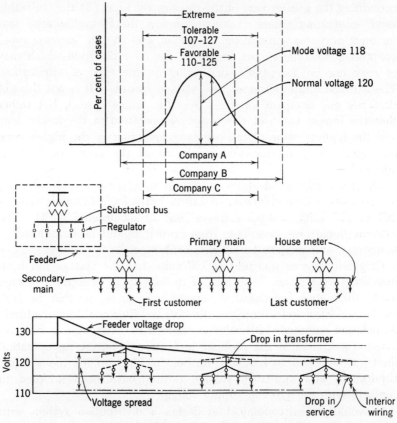

Figure 12-10. Illustrative representation of distribution voltage zones and voltage profile on a loaded circuit.

and he will receive high voltage. On the other hand, the heavily loaded last customer with maximum voltage drop between him and the supply point will receive low voltage. These relationships are illustrated in Figure 12-10.

The maximum drop of 10% usually is made up of 3% drop in the primary feeder between the first and last customers, 3% in the distribution transformer, 3% in the secondary mains, and 1% in the customer's service leads. If the first transformer always carries some

Power Distribution 231

load and the drop through it therefore is never less than, say, 2%, advantage will be taken of the fact to allow that much more drop in the primary.

Bus regulation can be used only where all the feeders leaving a substation are of about the same length and have the same character of load, as is generally the case in any given residential area. Such regulators merely raise or lower the voltage in response to the total load on the substation. They in no way change the distribution of voltage drops in the area. Where considerable difference exists between individual feeders, as regards either length or loading, individual feeder regulators are used. The regulators are mostly single-phase, two in three-wire, three-phase circuits and three in four-wire circuits.

Loads on distribution circuits are mostly single-phase, and some unbalance in the primaries is unavoidable. Three-phase regulators hold the voltage on one phase only, and the other two phases change correspondingly. Distribution transformers on those two phases will, therefore, have their voltage corrected when no correction is needed, or they may even get correction in the wrong direction. This is a disadvantage of three-phase regulators. If copper has to be added to the feeders, because of errors in correction, the extra copper may more than counterbalance the gain obtained by the use of one three-phase regulator instead of three single-phase regulators.

Induction regulators are equipped with contact-making voltmeters and compensators. The purpose of the compensator is to cause the regulator to hold a higher voltage at peak load than at light load. It consists of adjustable resistance and reactance elements in the circuit of the regulator which bias the contact-making voltmeter, depending on the value of current flowing through the regulator.

The problem of regulation in rural lines differs from that in urban districts. The lines being long, the voltage is usually higher (13,200 volts), and the load per mile is small; therefore, the costs must be kept down. The regulation tolerated is more likely to be 12 to 15% rather than 10%. In many cases copperweld conductor is used on spans of 300 to 600 ft. The transformers supply one or two customers and are usually not over 5 kva capacity. Where customers are 1000 ft apart, it is usually cheaper to provide separate transformers for each. The voltage regulators used on these lines are usually of the step-regulator type. They consist of a shunt or exciting transformer (two-winding or auto), a series transformer, and necessary accessories in one tank. The exciting transformer has several taps, one of which is selected by an arrangement of selector switch and breakers in

232   Principles of Electric Utility Engineering

response to signals from a contact-making voltmeter. These regulators are mounted on poles along the primary feeder line and buck or boost the voltage automatically by means of a contact-making voltmeter, as in the case of induction regulators.

Diversity

Diversity of a system was defined in Chapter 1 as the difference between the sum of the power demands in the system and the maximum coincident demand of the system, where "system" may be any part of the utility. There is diversity between residential consumers, between distribution transformers, between primary feeders, and so on. Diversity is the factor in a utility which makes possible its low rates. Without diversity the capacity of the installed equipment required to supply a given amount of energy would have to be three to four times its actual value.

Many studies have been made to determine the diversity in different parts of utility systems. The result is usually expressed as a diversity factor, that is, the sum of the maximum demands of a group of consumers divided by the coincident maximum demand of the group as a whole.

The loads tend to smooth out (DF = 1) as their number increases, so that the larger the system and the closer we come to the source of supply the better becomes the diversity factor. In a large utility system there is probably very little diversity between transmission circuits unless they supply loads of very different character—for example, large factories as against residential areas—or span different time zones.

Representative values of diversity factor between residences and a bulk supply substation are as follows:

Between residences	2–3
Between distribution transformers	1.2–1.5
Between primary feeders	1.1–1.2
Between subtransmission feeders	1.1–1.2

An example of diversity is given in Figure 12-11. It is assumed that the residences on a suburban avenue each have a frontage of 80 ft and a maximum demand in the form of lighting and appliances of 5 kva. With a diversity of 2.2 and using transformers of 25 kva, the houses can be grouped in lots of eleven, and a secondary conductor of No. 2 copper with a voltage drop of 2.5% is practicable.

If the primary feeder has forty such transformers, the sum of the peaks is $40 \times 25 = 1000$ kva, but with DF = 1.2 between transformers

Power Distribution

the substation transformer need have a capacity of only 833 kva; and so on back to source of supply. Without diversity, the peak on the station is $5 \times 11 \times 40 \times 6 \times 4 = 53{,}000$ kva; with diversity, it is 14,600 kva. The over-all diversity factor is therefore 3.63.

The loss of diversity in residential areas after an outage, especially in the evenings, can be troublesome. During such an outage, people

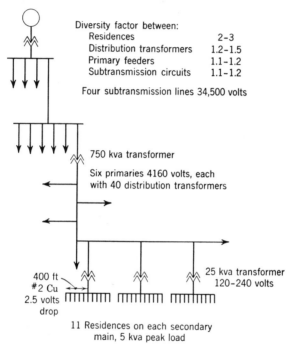

Figure 12-11. Example of load diversity from a group of residences to a bulk power source.

leave the light and radio switches closed, the refrigerators warm up, and the water heaters cool; consequently when service is restored all the load comes on at once. There is no diversity. Where automatic reclosers are used the inrush current may well cause them to trip out and make restoration of service quite difficult. About the only solution is to sectionalize the circuit and re-energize it in parts. This sectionalization can be done automatically.

Distribution in Commercial Buildings

Since the introduction of air conditioning in commercial buildings, the power demand has increased sharply. Without air conditioning the demand varies from 3 to 7 kw per 1000 sq ft of floor area, with a general average of 5 kw. Air conditioning adds another 3 to 5 kw. Since the buildings are usually in a congested part of town and take the form of height rather than expanse, it becomes difficult to

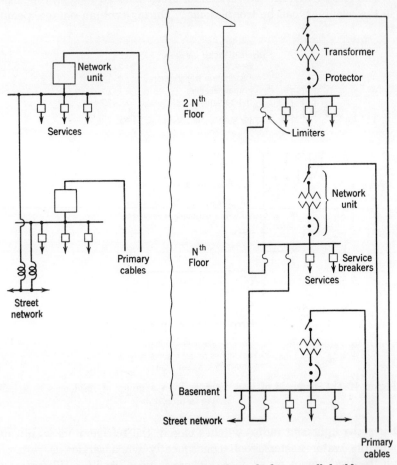

Figure 12-12. Illustration of a network applied to a tall building.

supply all the load from the street level. The network system, usually available in congested areas, lends itself well to such situations. The primary suppy is extended vertically up the building, and network units are located near the load centers (elevators, air conditioners, etc.) on the upper floors. The basement unit is usually tied into the regular street network as illustrated in Figure 12-12. Small building networks connected to two different primary feeders may sometimes

Power Distribution

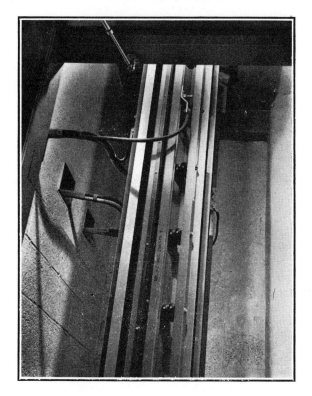

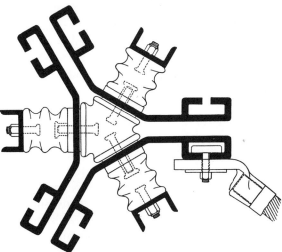

Figure 12-13. Secondary riser in a tall building. (Courtesy of Aluminum Company of America.)

form a relatively low impedance path between the supply circuits. If there exists a difference of voltage between the supply circuits, equalizing currents will flow through the low impedance path and cause one or the other of the network units to trip. In these circumstances it becomes necessary to install reactors or balancing coils between the network transformers and the secondary bus to prevent unnecessary operation of the protectors.

The primary voltage vertical riser cable to supply the transformers on the upper floors is of the paper-insulated, lead-covered type and frequently is of special design to prevent internal slippage. The low-voltage (208 volts) supply to intervening floors usually also is cable. There is on the market, however, a rigid riser bus of sector-shaped extruded aluminum which can be run up the building in a small shaft alongside the elevators, and to which at each floor the cables supplying that floor are connected as shown in Figure 12-13. This bus has a very low reactance and consequently low voltage drop, and permits greater spacing (more floors) between transformers. It greatly simplifies installation and maintenance.

Distribution in Industry

The largest individual customers of the utilities are the mining and manufacturing industries. The "large light and power" customers take, on the average, between 45 and 50% of the energy sold by the utilities, and they themselves generate about one fourth of that amount. In the 25-year period of 1925-1950, while the number of production workers increased 50% and the Federal Reserve Board's index of physical production doubled, the kilowatt-hours purchased by the large industrials from the utilities increased five times. There is no indication that the growth is likely to slow down in the forseeable future.

The manufacturing industries fall into two broad classifications: (1) producers' goods or capital goods and (2) consumers' goods. Capital goods comprise such products as machinery, girders, and bricks, which are used by others to produce consumers' goods or other producers' goods. Consumers' goods include all those articles used by the ultimate consumer such as clothes, food, ash trays. They are subdivided into three classes: (1) durable goods, such as pianos and automobiles, (2) semi-durable goods, such as shoes and clothing, and (3) perishable goods, such as foods and tobacco.

The foregoing distinctions have a marked influence on the nature and the layout of the manufacturing plant and the electrical equipment used. In a factory producing capital goods and durable goods, the equipment is more likely to be heavier and more permanent than in one producing perishable goods. Flexibility is probably less important; machinery, such as heavy boring mills, once located is not likely to be moved for years. Consumers' goods, on the other hand, change in style, and a factory producing such goods must be prepared to make the necessary adjustments. Flexibility here is of prime importance.

The choice of a power-distribution system for a large manufacturing establishment constitutes one of the most difficult problems the engineer has to solve. Only two factors are of interest to the owner of the factory—cost and continuity of service. In most cases power is a small part of his over-all costs and is looked upon as a necessary evil. He is therefore not inclined to put his capital into power-distibution equipment which will ensure better service when the same amount of money put into productive machinery will bring in greater returns as long as the power-supply system works. It is only when a power interruption occurs and production drops that the question is raised—would it be wise to bolster up the power-distribution system?

The problem that faces the factory owner is a difficult one—the balance between a real capital outlay and the probability of an outage. The real outlay for different kinds of distribution systems is easily figured. The real cost of an outage is something quite different. The factors that enter into the cost are numerous and rather vague. If the article being manufactured is a mass-production device sold off the shelf at retail, a shutdown of a few hours may not be serious. If the article is a custom-made piece of machinery for a specific customer, that customer may not be too happy if, as an excuse for delay, he is told the shop distribution system broke down. If the shop is a gun factory in time of war, such an excuse would certainly cause trouble. If the article is one manufactured seasonally, such as rubber shoes or summer hats, not only is continuous power supply essential during the period of production, but also there may be required periodic changes in the production set-up. In such cases a more expensive distribution system may prove more economical over the years because of its greater flexibility. The greater reliability in such cases is obtained at no extra over-all cost.

The general lay-out of the plant influences the choice. In an old factory with several floors, the wiring for the most desirable system may present difficulties, whereas in a new factory with all manufacturing on one level dependability of service may be obtainable at very little, if any, additional cost.

Small factories with a few relatively small machine tools are usually supplied with power at 240 volts from the delta secondary of a transformer on the pole in the street. One phase of the delta grounded at the middle point will supply lighting and fractional horsepower motors at 120 volts. In such a case there is no distribution problem beyond seeing that all the code regulations are met. All installations, large and small, must meet the standards of the National Electric Code of the fire underwriters. These standards represent minimum requirements of safety. They apply to types of equipment, wiring methods, circuit protection, grounding, conductor sizes, etc. Good service often requires something better than the minimum stipulated.

In large factories with several buildings distribution involves both primary feeders and utilization circuits. The primary voltage is usually 4160 or 13,200 volts. These voltages permit the use of 5000- and 15,000-volt switchgear, respectively, to full capacity with the least current and therefore the smallest conductor section. Supply at 2400 volts is desirable only if there are many very large motors which could be connected directly to the 2400-volt primaries. The choice between 4160 volts and 13,200 volts depends on the required rupturing ability of the breakers. If the rupturing requirement is less than 50,000 kva, the lower cost of the 5000-volt switchgear more than offsets the extra cost of the heavier conductors; but if the required rupturing ability is greater than 50,000 kva, the 5000-volt switchgear becomes almost as costly as the 15,000-volt switchgear, and the heavy conductors are more costly.

The secondary voltage is preferably 480 volts, unless a factory network is used, in which case the voltage is 120/208 or 265/460. The switchgear is of the 600-volt class. Induction motors with standard ratings up to 1000 hp and with a wide range of speeds are obtainable from all manufacturers as standard equipment at 440 volts. For network systems 220-volt motors are standard up to 300 hp, but much larger motors are obtainable. For instance, motors up to 1250 hp are operating on the New York 208-volt network system.

For the normal run of factories the load density for light and power run from 5 to 30 kw per 1000 sq ft, with the majority between 15 and 20 kw. Of this 10 to 20% may be for lighting, 4 kw per 1000 sq ft being a good average figure. The demand factor, that is, the ratio of the actual peak load to the installed capacity, varies from 25 to 90%. Machine shops have the lowest demand factors, 30 to 50%, and the continuous process mills, such as paper and textile mills, the highest, 70 to 90%.

The power supply within the factory can take many forms. The simplest consists of a single transformer with a number of radial feeders to the load centers. The most complex consists of two or more main feeders supplying network transformers at the load centers. Some years ago the distribution engineers of the Westinghouse Electric Corporation made a careful analysis of a large number of possible schemes and compared their relative merits on the basis of six major characteristics:

1. Investment—actual cost of all equipment and its installation.
2. Operation and maintenance costs—includes ability to inspect and repair all equipment with a minimum interruption to service.
3. Efficiency—includes the over-all system from bulk supply point or points to the utilization devices, for both heavy-load and light-load conditions.
4. Regulation—takes into account changes in voltage with load, whether slowly or suddenly applied.
5. Service continuity—measured in amount of load dropped with faults at different locations on the system.
6. Flexibility—the system's adaptability to changes in load conditions with minimum service interruption and cost.

Of the many schemes studied the following four, illustrated in Figures 12-14 to 12-17, have been selected as representative.

1. *Radial System*

The simple radial system is the one found in the majority of older plants. It uses a single substation, where power is received at the supply voltage and stepped down to the factory distribution voltage by one or more transformer banks. Energy may be metered on the high- or low-voltage side, depending on whether the utility or the factory owns the transformers. Feeders are run to the various load centers from the single substation bus.

Since the entire plant load is fed through a single substation, full advantage is taken of the diversity betwen load centers which permits a minimum amount of transformer capacity. However, the voltage regulation and efficiency are poor, the cost of feeders and switchgear is high, and flexibility is poor. Trouble in the substations shuts down the whole plant.

Some improvement is obtained if the utility high-voltage circuit is taken to several smaller substations nearer the load centers, instead of to one substation. This plan cuts down the cost of secondary distribution materially. The feeder from the utility terminates at a bus to permit the installation of a main breaker and metering equipment. The breaker on the incoming line may be the only one. In this case a fault in the high-voltage cables or transformers in the plant shuts down the whole plant. Where this hazard is undesirable, high-voltage breakers may be inserted in each feeder from the bus to the power centers. The transformers are usually three-phase, with an integrally

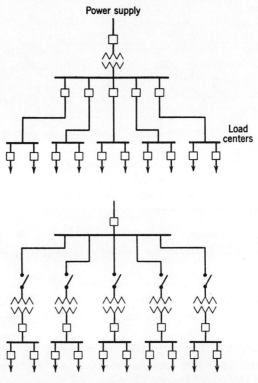

Figure 12-14. Factory distribution. Radial systems.

mounted primary switch and draw-out-type low-voltage switchgear, all bolted together to form a complete factory-built unit-type substation. The diversity is, of course, reduced so that the total kva transformer capacity required is larger than in the first type of substation described.

2. *Banked Secondary System*

This system uses a primary and secondary loop connection. Manually operated load-break switches are installed in the feeder, one at each transformer. A primary feeder or transformer fault shuts down the entire plant by tripping the main breaker, but service can be

Power Distribution 241

quickly restored after the faulty section has been isolated by opening two load-break switches and one transformer secondary breaker.

The secondary loop equalizes the load on all transformers, so that a careful fit of the transformer capacity to the load at any center is unnecessary. It generally permits the use of one transformer size

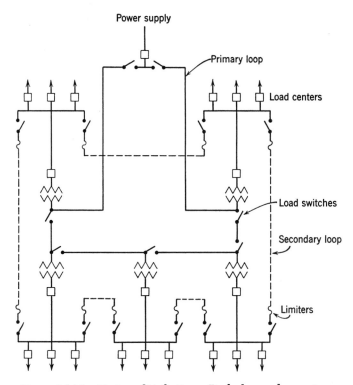

Figure 12-15. Factory distribution. Banked secondary system.

throughout the factory, thus facilitating the question of spare parts. It permits the load circuits to run from the load to the nearest load center. This fact reduces the length and cost of secondary conductors. Advantage is taken of the diversity because all transformers are connected in parallel; a saving in transformer capacity results. On the other hand, the fact that all transformers are in parallel results in a higher short-circuit current.

In both the foregoing schemes a cable fault shuts down the whole plant. Where this risk is not permissible two or more cables must be brought into the factory. In large plants such duplication may be necessary for other reasons than continuity of service. For them a so-called primary selective system or a secondary network should be considered.

3. Primary Selective System

Two primary cables are brought to each load center transformer. If one feeder is out of service, the other carries the load. Normally half the transformers are connected to each feeder. When a fault occurs on one feeder, there will be an interruption of service to the

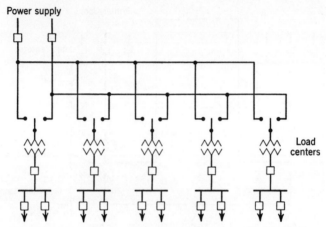

Figure 12-16. Factory distribution. Primary selective system.

transformers connected to it until they can be switched to the unaffected feeder. If a fault occurs in one of the transformers, its associated primary feeder breaker opens, and half the load in the plant is dropped, until the transformer is disconnected manually.

4. Secondary Network System

This is the same kind of network system that has been used for metropolitan distribution for many years. Power is fed to the load centers through network protectors, and each load bus is tied to the adjacent load bus through a secondary loop. There must be at least two primary feeders to make the system effective. If there is a fault in a primary feeder or transformer, the protectors of the units connected to that particular feeder open and isolate the faulted equipment. There will be no interruption to service because the load is carried by the neighboring transformers through the secondary loop. The loop also allows advantage to be taken of the diversity among the various load centers; it equalizes the loads on the transformers and usually allows a reduction in total secondary copper.

Power Distribution 243

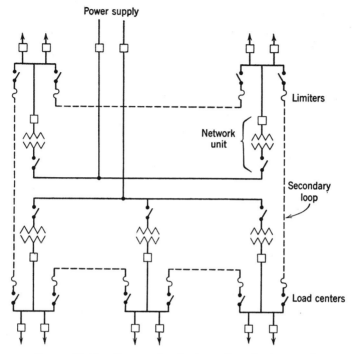

Figure 12-17. Factory distribution. Secondary network.

When these four schemes are evaluated in terms of the six characteristics previously given and the values are designated by the letters A to E, where $A = 91–100\%$ desirability, $B = 81–90\%$, $C = 71–80\%$, $D = 61–70\%$, $E = 51–60\%$, their relative standing is as shown in this table:

System described	1	2	3	4
Investment	D	D	B	E
Operation and maintenance	A	B	A	B
Efficiency	E	A	B	C
Voltage regulation	E	B	D	A
Service continuity	E	C	D	A
Flexibility	C	A	D	A

A study of this table indicates that in the matter of flexibility the banked secondary and the secondary network rate highest, as might be expected. This rating occurs because in these systems all the transformers are available to carry the load irrespective of where the load is added. Moreover, if regulation becomes troublesome, it is easy to cut the secondary ties halfway between two load centers and add another load center. Continuity of service is not so good with the banked secondary as with the network because in case of primary or transformer trouble there are no protectors to remove the fault automatically without a secondary interruption. Where uninterrupted service is essential, a secondary network is essential.

With the grading system used the ideal system would be one in which all items rated A or 90–100%. If the first two items are combined as "cost" and the last four as "performance," the following grading results:

	Cost	Performance
Conventional radial	85%	65%
Banked secondary	80%	92.5%
Primary selective	95%	75%
Secondary network	75%	95%

From this it is seen that from the point of view of cost the simple radial system is the most desirable, but it is the least desirable in performance. On the other hand, the secondary network shows the opposite result. Combinations and modifications of the extremes lie in between, and the choice in an actual application depends on the weight given to the various factors.

Where factory networks are used, the voltage is 120/208 or 265/460 volts. The lower voltage is suitable where a large part of the load is in the form of small motors. Lighting is connected from phase to neutral as in a metropolitan network. Where the higher voltage is used, the lighting in part takes the form of 265-volt high-intensity, high-bay, mercury-vapor lamps. However, 120-volt circuits (usually from small dry-type transformers) must also be provided for incandescent lights and hand tools.

Secondary feeders are usually of cable or busway. Busways are copper or aluminum bars supported on insulators in a ventilated enclosure suspended from the ceiling or from brackets on the wall. They are usually sold in 10-ft lengths with necessary accessories such as elbows, tees, cable tap boxes, bus plugs, and switches, so that connections to them can be made almost anywhere. This is the most flexible scheme available for supplying power to machine-tool motors and other electric equipment. Standard busways are available from 250 to 4000 amp and 600 volts. As loads are reduced along a busway, its capacity can be reduced correspondingly.

Where cable is used, it is usually rubber insulated. The conductor size must be large enough to satisfy two conditions: (1) current-

Power Distribution 245

carrying capacity must be adequate to prevent overheating, and (2) voltage drop must be within allowable limits. When the load, load factor, power factor, and ambient temperature are known, there is nothing complicated about calculating the voltage drop and size of conductor, but in general the calculation is unnecessary because the information can be obtained from the manufacturers' publications. It is preferable not to use a cable (single conductor) larger than 500,000 CM. Such a cable has a current-carrying capacity of about 400 amp continuous rating at 50°C copper temperature, assuming three conductors in one duct. Where the current is greater than 400 amp, it is preferable, although not always possible, to use two conductors in parallel rather than a single heavier cable. The smaller cables are easier to handle, and the reactance is lower.

The calculation of short-circuit current in the distribution system of a factory supplied by a public utility is the same as for utility circuits except that many simplifications are permissible. This is because almost invariably considerable reactance intervenes between the fault and the source of generation. The voltage used in the calculations is the name-plate voltage on the high side of the transformer supplying the plant. If it is assumed that this voltage is maintained during the fault, a factor of safety is automatically introduced into the results. If the reactance is used instead of the true impedance, another factor of safety is introduced.

Two values of short-circuit current interest the industrial user: the one-cycle value which determines the mechanical stresses and momentary ratings of apparatus, and a second value for determining the interrupting duty of the breakers. Both these values are obtained by applying a multiplier to the current value obtained by dividing the system reactance from the supply transformer to the point of fault into the system voltage to neutral, E/X.

Large factories have high-voltage circuits (2300 to 13,200 volts) as well as low-voltage circuits (below 600 volts). For the high-voltage circuits the multiplier for the 1-cycle value is 1.6 and that for determining the breaker duty varies from 1.0 for 8-cycle breakers to 1.4 for 2-cycle breakers. For low-voltage air circuit breakers only one multiplier is used, 1.25.

The foregoing statements rather oversimplify a complex problem. If the factory, in addition to purchasing power, generates power, which is frequently the case where process steam is used, the multipliers above may not apply. In low-voltage circuits (below 600 volts) the current evaluation of impedances is more important than in high-voltage circuits; a small impedance can materially affect the result.

Likewise the power fed into a short circuit by the induction motors cannot always be neglected. For these reasons it is advisable before applying breakers to study the breaker application data books put out by the various manufacturers to evaluate properly all the factors to be considered.

INDEX

Air conditioning, as it affects power demand, 233
Air preheater, in boiler plant, described, 40
American Gas and Electric Company, 195
Ammeter, surge-crest, 162
Arcing grounds, 121
Armature flux, described, 53
Arresters, and insulation of system, 173
 as they affect current wave, 172
 for rotating machines, 180
 illustration of, 177
 in fault control, 134
 in secondary system, 223
 location of, 178
 types of, 169
Ash sluicing, 76
 pump for, 78
Auxiliaries, for hydroelectric stations, 91
 for steam stations, 64

Base load, defined, 9
Basic impulse insulation level, defined, 173
 for high voltage, 195
Binary cycle, described, 42
Boiler control, 44
Boiler plants, description, 34
 fans for, 71
 improvement in, 65
 pumps for, 76
 stokers for, 71
Boiler rating, 44
Bonds, described, 17
Bonneville Administration, 195
Boston Edison Co., cost of Mystic station, 16

Brakes, for water-wheel generator, 91
Buchanan Dam plant, 81, 82
Buses, and stability, 201
 described, 125
 for substations, 217
 in fault control, 135

Cables, breakdown discussed, 110
 described, 108
 for primary system, 219
 high-voltage, 109
Capability, defined, 9
Capacitors, for rotating machines, 180
 series, 205
 shunt, 207
Capacity factor, 9, 13, 14
Capital, amount of investment, 15
 cost of, 17, 18
 described, 15
 how secured, 16
Circuit breakers, how they operate, 118
 in distribution system, 221, 223
 in fault control, 134, 136, 139, 156, 157
 ratings, 125
 reclosing time, 194
 types of, 122
Circuit reclosing, automatic, 194
Classifications, of utilities, 7
Clayton, J. M., 189
Coal, powdered, how handled, 68
Commercial buildings, power demand of, 234
Compressors, 78
Condensers, described, 51
 pumps for, 76
Conductors, configuration determined, 187

247

Index

Conductors, cost determined, 192
  for primary system, 219
  for secondary system, 223
  how chosen, 102, 185
  in network distribution, 226
  insulation for, 189
  loading limited, 183
  material of, 105
  spacing of, 107
Conowingo station, 79
Contactor, defined, 118
Contractors' fees, cost of, 192
Conveyors, for coal, 69
Cooper, Dexter, 29, 30
Corona, as it affects conductors, 185
  cause of, 106
Costs, classified and described, 15
  affected by demand, 1
  fixed charges discussed, 16
  production, discussed, 20
Counterpoises, 191
Crushers, for coal, 69
Cubicles, defined, 125

Damper windings, for water-wheel generators, 90
Deaerator, in boiler plant, 41
Demand, defined, 9
Depreciation, how determined, 18
Diablo station, 79
Direct-current transmission, advantages, 98
Distribution system, definition of, 7, 212
  protection for, 220
Diversity factor, defined, 9, 14
  discussed, 232
Dust, control of, 44

Economizer, in boiler plant, 40
Economy in steam-generating stations, 40, 41, 42
Electric utilities, organization, 3
  profit, 22
  property, 5
  rates, 8
  standards of service, 2
  stock described, 17
  taxes, 20
Engineering department, subdivisions in, 4

Engineering supervision, cost of, 192
Exciters, for generators, 59, 61, 62, 201

Fans, for boilers, 71
Faults, bus, 150
  machine, 148
  transformer, 150
  transmission-line, 151
  types of, 141
  unbalanced, 147
Federal Power Commission, definitions of transmission by, 96
  regulations of, 6
Feeder circuits, in fault control, 135
Feedwater heaters, in boiler plant, 40, 43, 44
Feedwater make-up, pump for, 76, 78
Fifteen Mile Falls station, 79
Fire protection, pumps for, 76
Franchise, duties and rights of, 2
Francis type turbine, 82
Frequencies, control, 209
  converters, 100
  standards, 97, 99
Fuel, amount available, 25
  amount used, 20
  artificially made, 31
Fulchronograph, 163
Furnace, in boiler plant, 35
Fuse links, 221

Gas turbines, use discussed, 27
General Electric Company, 42
Generators, and stability, 201, 208
  described, 52
  double-winding, 146
  exciters for, 59, 61, 62, 201
  for auxiliary power, 65, 67, 78
  grounding, 157
  hydrogen-cooled, 55
  induction, 82
  inner-cooled, 56
  Marx, for surge testing, 174
  output limited, 55
  parallel operation, 62
  reactance discussed, 57
  size discussed, 62
  vibration discussed, 58
  water-wheel, types, 88
George, E. E., 181

Index

Governors, for frequency control, 209
  for water wheel, 87
Grand Coulee Dam, 79, 81
Grounding, cost of rods, 154
  low-voltage systems, 156
  of generators, 157
  system, described, 154
  types described, 152

Harder, E. L., 189
Hartford Electric Company, 33
Hoover Dam, 79, 81, 123, 190
House cranes, 78
Hydroelectric stations, automatic, 82
  cost of power, 93
  cost of transmission studied, 182
  economic justification for, 93, 94
  government-built, 79
  private, 79
  pump storage plants, 81
  types, 79
Hydrogen cooling, of generators, 55

Incremental loading, 208
Incremental rate value, 208
Industries, distribution of power in, 236
  forms of power supply, 239
Insulation, amount necessary on conductors, 189
  levels, 173
  types, 175
Insulation coordination, 173
Isokeraunic level, 189

Kaplan type turbine, 82
Kelvin's law, 102, 182
Klydonograph, 161

Labor, types of, 21
Lamme, B. G., 203
Leeds and Northrup tuned-bridge controller, 211
Lightning, as it affects transmission lines, 188
  explanation of, 160
  protection against, 167
  standard wave established, 172
  stroke currents measured, 161, 163
  types of, 166
Load curve, defined, 9, 10

Load diversity, defined, 9
Load duration curve, defined, 9, 11
Load factor, defined, 9, 13
Loading, amount allowed, 9, 183
Loading schedules, and stability, 208
Los Angeles Department of Water and Power, 191

Machines, rotating, lightning protection for, 180
  synchronous, as they affect stability, 201
McIndoes plant, 80
Manhattan Elevated Railway Company, 33
Martin tables, 185
Marx, Dr. Emil, 174
Metropolitan Water District of Southern California, 190
Mitchell station, 16
Monopoly, competition in spite of, 2
  reasons for, 1
Motors, squirrel-cage, 69, 71, 77, 93
Mystic station, 16

National Electric Code, 238
Network analyzer, 202
Network distribution, 223
Nuclear energy, use discussed, 28

Osage station, 79
Oscillographs, 161

Pelton type turbine, 82
Pensacola Dam, 81
Petersen coil, 156
Peterson, E. L., 186
Peterson, W. S., 193
Phase segregation, in fault control, 146
Philips station, 16
Pierce, R. E., 181
Plant factor, 9
Power, auxiliary supply in stations, 64
  cost figured, 93
  normal supply in stations, 65
Power-angle diagram, 197
Power circle diagram, 203
Primary cutout, 220
Primary distribution circuits, 219

Index

Public Service Electric Gas Company of N. J., 16
Pulverizer, for coal, 68
Pumps, described in stations, 76

Reactance, and stability, 201, 205
  in valuing short-circuit current, 141
  of generators, 57, 139, 157
  of transformers, 140
Reactors, in fault control, 134, 135, 138, 157
Reclosers, in distribution system, 221
Reheat, in boiler plants, 40–41, 43, 49
Relays, in fault control, 134, 148
  microwave, 152
  purpose of, 127
  types of, 128 ff.
Resistor, in fault control, 157, 158
Right-of-way, cost of, 193
Risers, for commercial buildings, 236
Rock Island plant, 80
Rocky River plant, 81, 82
Rod gaps, 176
Rotors, in water-wheel generators, 90

Sag, in determining span, 183
St. Clair, H. P., 186
Salt River Valley Water Users Association, 97
Secondary distribution system, 221
Sectionalizers, in distribution system, 221
Sewaren station, 16
Shale oil, availability discussed, 32
Short-circuit current, how determined, 141
  in industry, 245
Short-circuit ratio, defined, 53, 55
Slag screens, in boiler plant, 40
Slepian, Dr. Joseph, 119
Solar heat, use discussed, 29
Southern California Edison Company, 191
Span, how chosen, 183
Stability, how to maintain in system, 59
  illustrated, 197
  types of, 197
State utility commissions, 7
Steam electric stations, cost of power, 93

Steam electric stations, origin, 33
  parts of, 33
  supplant hydroelectric power, 79
Steam turbines, governors adjusted, 62
  ratings, 50
  types of, 45
Steel towers, 183
  weight determined, 193
Steinberg and Smith, *Economy Loading of Power Plants*, 208
Stokers, where used, 71
Substations, size of, 217
Subtransmission circuits, 212
Subtransmission lines, 215
Superheater, in boiler plant, 40
Superposition, in turbines, 48
Supervisory control equipment, 126
Supply circuits, in fault control, 135
Surge impedance, 186
Switchboards, described, 125
  power for, 126
Switchgear, defined, 117
  metal-enclosed, 125
Switching surges, 121
Switch yard, 111
System voltage recovery rate, 119

Thomas charts, 185
Tidal power, use discussed, 29
Tie-line bias frequency control, 210
Transformers, current, used with relays, 127
  designs of, 111
  distribution, in secondary system, 222
  for auxiliary power, 65, 67, 78
  insulation for, 176, 178
  network described, 225
  noise in, 117
  overload of, 117
  potential, used with relays, 127
Transmission, definition of terms by F. P. C., 96
  equipment, 111
  remote control for, 126
Transmission lines, amount of power on determined, 186
  cost determined, 192
  high voltage problems, 195
  lightning effects, 165, 188
  purposes of, 96

Index

Transmission system, cost of studied, 181
  definition of, 7
Turbines, *see* Steam turbines, Water turbines

Underground distribution, 224
Unit type stations, 65

Ventilation, for water-wheel generators, 90
Voltage, definitions, 228
  for factories, 238

Voltage, for residential service, 212
  how determined, 101
  regulation discussed, 229
  standard, 229

Warren pendulum clock controller, 211
Water power, amount of in utility, 24
Water turbines, how designed, 84
  types, 82
West Penn Power Company, 16
Wilson Dam, 81
Wind power, use discussed, 31
Wood poles, 183, 215

HETERICK MEMORIAL LIBRARY
621.31 P88p
Powel, Charles A/Principles of electric   onuu

3 5111 00136 2510

Heterick Memorial Library
Ohio Northern University
Ada, Ohio 45810